The OXFORD Illustrated Dictionary of AUSTRALIAN HISTORY

The OXFORD Illustrated Dictionary of AUSTRALIAN HISTORY

JAN BASSETT

Rock Valley College
Educational Resources
Center

Melbourne
OXFORD UNIVERSITY PRESS
Oxford Auckland New York

OXFORD UNIVERSITY PRESS AUSTRALIA

Oxford   New York   Toronto
Delhi Bombay Calcutta Madras Karachi
Kuala Lumpur Singapore Hong Kong Tokyo
Nairobi Dar es Salaam Cape Town
Melbourne Auckland Madrid
and associated companies in
Berlin Ibadan

OXFORD is a trade mark of Oxford University Press

© Jan Bassett 1993
First published 1993

This book is copyright. Apart from any fair
dealing for the purposes of private study,
research, criticism or review as permitted under
the Copyright Act, no part may be reproduced,
stored in a retrieval system, or transmitted, in
any form or by any means, electronic, mechanical,
photocopying, recording, or otherwise without
prior written permission. Enquiries to be made to
Oxford University Press.

Copying for educational purposes
Where copies of part or the whole of the book are
made under section 53B or section 53D of the Act,
the law requires that records of such copying be
kept. In such cases the copyright owner is
entitled to claim payment.

National Library of Australia
Cataloguing-in-Publication data:

Bassett, Jan, 1953– .
The Oxford illustrated dictionary of Australian history.
ISBN 0 19 553243 0.
1. Australia—History—Dictionaries. I. Title.
994.003

Edited by Nan McNab
Designed by Sandra Nobes
Typeset by Solo Typesetting, South Australia
Printed in Hong Kong by Nordica Printing Co Ltd
Published by Oxford University Press,
253 Normanby Road, South Melbourne, Australia

PREFACE

This book contains entries on people, institutions, places, ideas, movements, events, artefacts, and documents generally considered to be of significance in Australian history. Some entries concerning international events, with emphasis upon Australian involvement, are also included.

The entries are arranged in alphabetical order of their headwords. Cross-references are denoted by asterisks preceding the headwords of other entries in which relevant information may be found. They are not usually given for the names of states and capital cities, nor for frequently used terms such as the First World War. In the captions, relevant headwords in bold type relate the illustrations to specific entries.

ABORIGINAL FLAG Flag of the Australian Aboriginal people. The top half is black, representing the Aboriginal people; the bottom red, representing the earth; with a solid yellow circle in the centre, representing the sun. Harold Thomas, an Aranda elder from Central Australia, designed the flag, which was apparently first flown at *land rights demonstrations in Adelaide during the late 1960s. Aborigines at the tent embassy in Canberra in 1972 decided that they should fly an Aboriginal flag; a number of designs were considered, and Thomas's was chosen. It is not a registered flag.

ABORIGINES Term used to describe the original inhabitants of a country, as opposed to colonists. Australian Aborigines are believed to have come from South-East Asia to Australia, then part of a much larger land mass that included *New Guinea, more than 40 000 years ago. Migrations probably occurred over thousands of years. Most prehistorians now believe that there were two human types in prehistoric Australia who merged to form the modern Aboriginal race. The Australian Aborigines were semi-nomadic hunters and gatherers. Their systems of belief emphasized the Dreamtime and the spiritual significance of the land. Some 300 different languages were spoken. White settlement, which began in 1788, proved disastrous for the Aborigines. The Aboriginal population in 1788 has been estimated at about 300 000, but recent research suggests that it may have been much higher. Within a hundred years the population had declined to about 50 000, and virtually all of the Tasmanian Aborigines were dead. The main reasons suggested for this decline include loss of land, adoption of European habits such as drinking alcohol, effects of European diseases such as *smallpox, influenza, respiratory and venereal diseases, declining birth-rates, and violence between Europeans and Aborigines. Such violence (for example, at the Battle of *Pinjarra, at *Myall Creek, and that caused by the *native police) led to deaths on both sides. It is estimated that about 2500 Whites and about 20 000 Aborigines were killed in direct conflict. During the nineteenth and twentieth centuries, official White policies toward Aborigines have included, at various times, 'protection', segregation, assimilation, and integration. Aborigines have faced legal, political, economic, and social discrimination. A referendum proposing that Aborigines should be included in the census count, and that the Commonwealth Government should be given the power to legislate for Aborigines, was passed overwhelmingly in 1967. Most 'protective' and discriminatory legislation has been repealed. The *land rights movement, which began in the 1960s, continued to be significant in the 1990s.

ACT *Australian Capital Territory

ACTU *Australian Council of Trade Unions

ADELAIDE, the capital city of *South Australia, lies between the Mount Lofty Ranges in the east and Gulf St Vincent in the west. *Flinders explored the coast in 1801 and 1802, Collet Barker the land in 1831. The Colonization Commission instructed *Light, South Australia's first surveyor-general, to select the site for the new province's capital. He did so in 1836. His choice caused some controversy. He also planned the capital (which was named after Queen

Charles Bayliss's Group of Aborigines at Chowilla Station, Lower Murray, SA, one of a number of photographs of **Aborigines** taken by Bayliss in the latter part of the nineteenth century. (National Gallery of Victoria)

Adelaide), surrounding the central area with a wide band of parklands. Adelaide's municipal corporation, the first in Australia, was established in 1840. The population of Adelaide was 546 in 1836, 118 000 in 1886, and 1 023 700 in 1988.

'ADVANCE AUSTRALIA FAIR', the Australian national anthem since 1984, was composed by Peter Dodds McCormick, a Scot, in about 1878. It was first played at a concert in Sydney, arranged by the Highland Society of New South Wales, on St Andrew's Day, 30 November 1878. The revised words of the first stanza are:

> *Australians all let us rejoice,*
> *For we are young and free;*
> *We've golden soil and wealth for toil;*
> *Our home is girt by sea;*
> *Our land abounds in nature's gifts*
> *Of beauty rich and rare;*
> *In hist'ry's page let every stage*
> *Advance Australia Fair.*
> *In joyful strains then let us sing*
> *Advance Australia Fair.*

AGE *Syme, David

AIF *Australian Imperial Force

AIR FORCE *Royal Australian Air Force

ALBANY Town situated on King George Sound in Western Australia. George Vancouver visited and named the sound (originally King George the Third's Sound) in 1791. The British, motivated partly by fears that the French might settle in the area, established the first White settlement in the western part of Australia there in 1826. Major Edmund Lockyer, who led the first settlers, a small group of troops and convicts from New South Wales, named the settlement Frederickstown (after Frederick, Duke of York and Albany). It became known officially as Albany in 1832. The troops and convicts were withdrawn in 1831, the year in which the recently founded colony of Western Australia took responsibility for the settlement. Albany was Western Australia's most important port for many years.

ALFRED, PRINCE, DUKE OF EDINBURGH (1844–1900), was the subject of an unsuccessful assassination attempt during his tour of the Australian colonies between October 1867 and April 1868, the first made by any member of the British royal family. Irish-born Henry James O'Farrell (1833–68) shot and wounded Prince Alfred, Queen Victoria's second son, during a public picnic at Clontarf near Sydney on 12 March 1868. A bystander was also shot and wounded. The apparently insane O'Farrell claimed that he had been influenced by Melbourne Fenians, members of the Irish Republican Brotherhood, a revolutionary nationalist organization, but later withdrew this claim. The New South Wales Government, believing that the attempt had been part of a Fenian conspiracy, passed a *Treason Felony Act* on 19 March 1868, which contained strict provisions including the prohibition of disrespectful language. O'Farrell was tried, found guilty, and, despite the Prince's efforts to save his life, hanged at Darlinghurst Gaol on 21 April 1868. Prince Alfred recovered rapidly.

ALICE SPRINGS Town in the Northern Territory, close to the centre of Australia. The original springs on the Todd River, some kilometres from the present-day town, were first discovered by Whites in 1871, during preparation for construction of the *Overland Telegraph Line, and were named after the wife of Charles Todd, Superintendent of Telegraphs. The telegraph station was also named Alice Springs. The town, which was surveyed in 1888, was originally named Stuart (after John McDouall *Stuart), but became popularly known as Alice Springs; its name was officially changed in 1933. Alice Springs was the administrative centre of Central Australia when the latter was a separate territory between 1926 and 1931. The town was connected by rail with Port Augusta in the south in 1929, and by a bitumen road to Darwin in the north (a distance of more than 1500 kilometres) during the Second World War. The Joint Defence Space Research Facility at Pine Gap, near Alice Springs, was established after the signing of an agreement between the Australian and United States governments in 1966. Recently Alice Springs has been a centre for tourist, cattle, and mining industries.

ALL FOR AUSTRALIA LEAGUE *United Australia Party

'ALL THE WAY WITH LBJ' *Vietnam War

ALP *Australian Labor Party

ANA *Australian Natives' Association

ANGAS, GEORGE FIFE *South Australian Company

ANGLO-JAPANESE ALLIANCE Treaty between Britain and Japan, first signed in

1902 and renewed in 1905 and 1911. The treaty, the initial purpose of which was to prevent Russian expansion in the East, was a factor leading to Japan's involvement in the First World War. It was ended at the Washington Conference, during which several treaties were signed, including one between the USA, Britain, Japan, and France in 1921.

ANGRY PENGUINS The name given to the members of an Australian modernist movement of the 1940s, founded and led by the poet and writer Max Harris (1921–). The movement arose partly as a reaction against the nationalism of the *Jindyworobaks, in which Harris had originally shown some interest. The name came from words in one of Harris's poems, used by the poet Sir Charles Rischbieth Jury at an Adelaide literary dinner party to describe Harris and a number of other radical young poets. It was also used as the title for a quarterly cultural journal, mostly edited by Harris and John Reed, which was published between 1940 and 1946, the first two issues by the Adelaide University Arts Association, the later ones independently by Harris and Reed in Melbourne. The Angry Penguins, who wanted to link Australian writers and artists to the European modernist movement, included writers such as Peter Cowan and Harry Roskolenko, painters such as Sidney *Nolan, Albert Tucker, John Perceval, Joy Hester, Danila Vassilief, and Arthur Boyd, and musicians such as the jazz figures Graham and Roger Bell. The *Ern Malley hoax helped to bring about the demise of the journal, but the ideas of the Angry Penguins continued to have some influence. The work of the artists has proved to be the most enduring.

ANSETT AIRLINES OF AUSTRALIA One of Australia's two main domestic airlines for many years, the other being the government-owned *Trans Australia Airlines (later Australian Airlines). Reginald (later Sir Reginald) Myles Ansett founded Ansett Roadways in Victoria in 1931. He began Ansett Airways, with a Hamilton-Melbourne air service relying on one aircraft, in 1936. The size of the fleet and the extent of the services increased in the next few years. Ansett merged his Pioneer Tours, a tourist coach service, and Ansett Airways to form Ansett Transport Industries in 1946. The latter bought Australian National Airways (ANA), a major competitor, in 1957 to form Ansett–ANA, which later became Ansett Airlines of Australia. A group of other companies took over Ansett Transport Industries in 1981.

ANSTO *Australian Atomic Energy Commission

ANTARCTIC TREATY Treaty signed in 1959 by Argentina, Australia, Belgium, Chile, France, Japan, New Zealand, Norway, South Africa, the United Kingdom, and the United States of America. It came into effect in 1961 and other countries later acceded to it. The Treaty reserves Antarctica for peaceful purposes, pledges international scientific co-operation, and prohibits nuclear explosions. It does not affect prior territorial claims, of which Australia has made the largest, the *Australian Antarctic Territory, nor can activities during its term be used to support claims.

ANZAC Name derived from the initials of the Australian and New Zealand Army Corps, which fought in the First World War. It is reputed to have been

ANZAC DAY

Sydney Nolan's painting, Kenneth Soldier *(1958), of a man wearing the uniform of the Light Horse, is one of the many works by Australian writers and artists to have been inspired by the* **Anzac legend**. *(Australian War Memorial)*

coined by clerks at General Birdwood's headquarters in Cairo and was in use by early 1915. The place where the Anzacs landed on the Gallipoli peninsula on 25 April 1915 soon became known as Anzac Cove. Originally the name was applied to those members of the corps who took part in the *Gallipoli campaign, but eventually it was applied to all Australian and New Zealand troops who fought in the First World War.

ANZAC DAY (25 April) commemorates the Anzac landing on the Gallipoli peninsula on 25 April 1915, during the *Gallipoli campaign. It has been observed every year since 1916, becoming a holiday in one state after another from about the mid-1920s. Its function has been broadened to commemorate the dead of other wars. Anzac Day rituals include dawn services, marches, wreath-laying, and speeches.

ANZAC LEGEND A significant theme in twentieth-century Australian history. It rests upon the assumption that Australia 'came of age' as a nation when the *Anzacs landed on Gallipoli on 25 April 1915. Ellis Ashmead Bartlett (a British war correspondent), C. E. W. *Bean, the *Australian War Memorial, and the *Returned Services League are among those people and institutions credited with helping to develop and maintain the Anzac legend.

ANZAC PACT Security treaty between Australia and New Zealand, signed in 1944. It was also known as the Australia–New Zealand Agreement, and in New Zealand as the Canberra Pact. The South Pacific Commission was established under it in 1947. The Anzac Pact was virtually superseded by the *ANZUS Pact in 1951.

ANZAM Popular name given to an arrangement by which Australia, New Zealand and Britain stationed defence forces in Malaya and Singapore. The arrangement began during the *Malayan Emergency, and continued after Malaya became independent in 1957, and after Malaysia was created in 1963. ANZAM was superseded by the Five Power Defence Arrangement in 1971.

ANZUS PACT Security treaty between Australia, *New Zealand and the USA, signed at San Francisco in 1951 and ratified by Australia in 1952. It states that each party recognizes:

that an armed attack in the Pacific Area on any of the Parties would be dangerous to its own peace and safety and declares that it would act to meet the common danger in accordance with its constitutional processes.

Related agreements have allowed the USA to establish bases in Australia, for example at North West Cape, at Pine Gap, and in the Woomera area. New Zealand effectively withdrew from the Pact following its 1984 decision to ban nuclear vessels (including American ones) from its ports.

ARCHIBALD, JULES FRANÇOIS (1856–1919), journalist, was born at Kildare, near Geelong, Victoria. He was baptized John Feltham, but later adopted the names Jules François. His father was a police sergeant. Archibald attended schools in Warrnambool before becoming an apprentice in the printery of the *Warrnambool Examiner* in 1870. From 1872 he worked in the printery of the *Warrnambool Standard*, and wrote some pieces for the *Hamilton Spectator* and the *Port Fairy Gazette*. In 1874 Archibald moved to Melbourne, where he worked as a stone-hand on the *Herald* and as a reporter on the *Echo* and then on the *Daily Telegraph*. He held several clerical positions, with the Victorian Education Department (1876–8); with a Queensland engineering firm; and with the *Evening News* in Sydney, where he became a reporter. Archibald and John Haynes, a fellow journalist on the *Evening News*, founded the *Bulletin* in 1880. After being jailed for failure to pay costs in a libel action, Archibald and Haynes became employees of William Henry Traill, who took over as proprietor and editor of the *Bulletin* in 1882. Archibald visited England, France, and the USA in 1883. When Traill sold out, Archibald and William McLeod took over as proprietors of the *Bulletin* in 1886 and Archibald was its editor until 1902. He spent some time in the Callan Park Asylum between 1906 and 1910. Archibald eventually sold his share in the *Bulletin* in 1914. He worked on *Smith's Weekly* for some months in 1919 before his final illness. He died in Sydney. He left an annual prize for portrait painting.

ARMY Britain provided troops for garrison duties in Australia between 1788 and 1870. The first were marines with the *First Fleet; then came the *New South Wales Corps, and finally members of various regiments. These troops were mainly used to control groups such as convicts during the *transportation period, *diggers during the *gold-rushes, and *bushrangers. They were used, for example, to suppress the *Castle Hill Uprising, the *Eureka Rebellion, and the *Lambing Flat Riots. All of the Australian colonies began to raise local forces during the mid-nineteenth century (there had been some earlier short-lived attempts to do so in New South Wales and South Australia), and they maintained them, in some cases intermittently, for the rest of that century. Some Australian colonists went individually to the *New Zealand Wars (formerly known as the Maori Wars), to which British armed forces were sent. A *Sudan Contingent was formed. Australians were also involved in the *Boer War. Following federation in 1901, the newly created Commonwealth Government assumed responsibility for defence, and so the various colonial forces were amalgamated. The Royal Military College at Duntroon was founded in 1911. During the First World War the *Australian Imperial Force was formed, and a second Australian Imperial Force was formed during the Second World War. Women, other than those in the medical services, were not accepted into the army until 1941. Since the end of the Second World War, Australian troops have been sent to various conflicts, including the *Korean War, the *Malayan Emergency, and the *Vietnam War. Various forms of *conscription (sometimes called national service) have been used at times to supplement numbers in the army. Australia's other armed forces are the *Royal Australian Navy and *Royal Australian Air Force.

ARTHUR, SIR GEORGE (1784–1854), soldier and colonial administrator, was born at Plymouth, England. He joined the army in 1804, and subsequently served in several countries. He administered British Honduras from 1814 to 1822, and was Lieutenant-Governor of *Van Diemen's Land from 1824 to 1836. During this time, Van Diemen's Land was separated administratively from New South Wales (1825); large land grants were given out, notably to the *Van Diemen's Land Company; the *Black Line was formed, and *Port Arthur was established. Arthur, a strong believer in transportation as a punishment, greatly improved the administration of the convict system, and the colony flourished under his rule. He went on to become Lieutenant-Governor of Upper Canada (1837–41) and Governor of Bombay (1842–6). He died in London. His publications include *Observations upon Secondary Punishment* (1833) and *Defence of Transportation* (1835).

ASEAN *Association of South-East Asian Nations

ASHES, THE *Cricket

ASIO *Australian Security Intelligence Organization

AUSTRALIA DAY
SATURDAY, JANUARY 26th, 1918.

In celebration of the 130th Anniversary of the Settlement of Australia, the London Branch of the

Australian Natives' Association

WILL HOLD A

CORROBOREE

AT 8 p.m. IN THE

CONNAUGHT ROOMS,

GREAT QUEEN STREET, off Kingsway.

CONCERT in the Crown Room
DANCING in the Grand Hall
REFRESHMENTS in the Drawing Room
CONVERSATION in the Edinburgh Room
SHELTER (if required) in the Basement

TICKETS (inclusive) **5/-** EACH.

Each member of the Australian Forces who purchases a 5/- ticket is entitled to a Lady's Complimentary Ticket on application at the Anzac Club & Buffet, 94, Victoria Street.

"And the Night shall be filled with Music,
And the cares that infest the Day,
Shall fold their tents, like the Arabs,
And as silently steal away."

R. BRISBANE CURD, Hon. Sec.,
94, Victoria Street, S.W. 1.

A pamphlet advertising an Australia Day *function organized by the London branch of the Australian Natives' Association (ANA) in 1918. Like many other ANA activities, it borrowed from Aboriginal ideas and customs. (Australian Natives' Association)*

ASSIGNMENT *Transportation

ASSOCIATION OF SOUTH-EAST ASIAN NATIONS (ASEAN) Alliance formed in Bangkok in 1967. Its original members, Indonesia, Malaysia, the Philippines, Singapore, and Thailand, were joined by Brunei Darussalam in 1984. Its aims are to foster stability and economic co-operation in South-East Asia.

AUSTRALIA DAY (26 January) commemorates the landing of *Phillip and the *First Fleet at Sydney Cove in 1788. It was first declared a holiday in 1838. In the past, different states have given it different names, such as Anniversary Day and Foundation Day. Recently, some states have held a public holiday for it on the following Monday, others on the actual day.

AUSTRALIA FELIX was the early name used for the Western District of Victoria. In 1836 Thomas *Mitchell so named the area (*felix* being Latin for happy) because of its beauty and fertility.

AUSTRALIA FIRST MOVEMENT A short-lived organization formed in Sydney in 1941, with P. R. Stephensen as its president. It expounded extreme nationalist and neo-fascist views. Four people linked with the organization were arrested on charges of conspiring to assist Japan in Western Australia in 1942. Two were convicted, imprisoned, and later interned; the others were acquitted, but still were interned. Sixteen men, mostly members, were interned in New South Wales in 1942. (Two other people, one a member, the other an ex-member, were also interned in New South Wales and Victoria, but other factors were said to be involved in their cases.) Most of the sixteen were released in 1942, two in 1944, and Stephensen in 1945. Mr Justice Clyne, who conducted a commission of inquiry (1944–5) into the affair, found that the four Western Australians had been guilty, that the detention of eight of the sixteen had been unjustified, and that they deserved financial compensation (which was paid), that there was fair opportunity for the sixteen to appeal, that restrictions imposed after release were fair, and that no further action should be taken.

AUSTRALIAN AGRICULTURAL COMPANY An agricultural and pastoral organization formed by a London syndicate in 1824. Its stated aim in its charter was 'the Cultivation and Improvement of the Waste Lands in the Colony of New South Wales'. Its first agent, Robert Dawson, began operations at Port Stephens, north of *Newcastle, in 1826. Its activities have included raising and breeding stock, leasing and selling land, and coal-mining.

AUSTRALIAN AIRLINES *Trans Australia Airlines

AUSTRALIAN ANTARCTIC TERRITORY A large area of Antarctica claimed and administered by Australia as an external territory. It covers the islands and territories between 45°E and 136°E, and 142°E and 160°E. The area was claimed, following Antarctic exploration by *Mawson and other Australians, by a British order-in-council in 1933. The *Australian Antarctic Territory Acceptance Act* 1933 was ratified in 1936. Australia conducts scientific and other research in Antarctica, and bases were established at Mawson in 1954, Davis in 1957, and Casey in 1969. Australia's claim to the Australian Antarctic Territory and other such claims, were suspended under the provisions of the *Antarctic Treaty.

AUSTRALIAN ATOMIC ENERGY COMMISSION (AAEC) Statutory authority established under the *Atomic Energy Act* 1953. Its activities included the control of uranium mining at *Rum Jungle, the establishment of uranium mining at Mary Kathleen and elsewhere, and the creation of the Lucas Heights research centre near Sydney. It was superseded by the Australian Nuclear Science and Technology Organization (ANSTO) in 1987.

AUSTRALIAN BALLET *Australian Elizabethan Theatre Trust; *Borovansky, Edouard

AUSTRALIAN BALLOT *Secret ballot

AUSTRALIAN BROADCASTING COMMISSION (ABC) Statutory authority established in 1932, replacing the Australian Broadcasting Company, and originally paid for by listeners' licence fees. It began broadcasting with eight capital-city and four regional radio stations in 1932, founded studio orchestras throughout Australia and began to hold subscription concerts in the 1930s, and later established symphony orchestras in the state capitals. Radio Australia, an overseas radio broadcasting service, began in 1939. In 1948 an annual appropriation from the Commonwealth Parliament was introduced to pay most of the Commission's costs. Its television broadcasting began with channels in Sydney and Melbourne in 1956, followed by others elsewhere in later years. Colour television broadcasting began in 1975, and FM (frequency modulation) radio broadcasting in 1976. The Commission was renamed the Australian Broadcasting Corporation and restructured in 1983.

AUSTRALIAN CAPITAL TERRITORY (ACT) A federal territory in which *Canberra, Australia's national capital, is situated. A site in the Yass–Canberra district of New South Wales was chosen in 1908, and transferred to the Commonwealth in 1911. A small area of land at Jervis Bay on the New South Wales coast was included, in the expectation that it would become a federal port; this area was increased in 1915.

AUSTRALIAN COLONIES' GOVERNMENT ACT (1850) is the usual name given to a British Act of Parliament that granted a limited form of self-government to *Victoria, *New South Wales, *South Australia, and *Van Diemen's Land. Its actual name was *Act for the better Government of Her Majesty's Australian Colonies*. It separated the *Port Phillip District (renaming it Victoria) from New South Wales, and granted legislative councils (two-thirds of whose members were to be elected, one-third nominated) to Victoria, South Australia and Van Diemen's Land. New South Wales had already been granted such a legislative council in 1842, with stricter franchise requirements than those imposed elsewhere. The legislative councils could make some laws, change their franchise and membership qualifications, and (subject to royal assent) split into upper and lower houses. Provision was made for similar measures in Western Australia.

AUSTRALIAN COUNCIL OF TRADE UNIONS (ACTU) The main organization governing Australian trade unions. The Australasian (from 1943, Australian) Council of Trade Unions was formed in 1927. It has played, and continues to play, a significant role in Australian industrial and political affairs. In 1992 there were 148 trade unions affiliated with the ACTU.

AUSTRALIAN COUNTRY PARTY *National Party

AUSTRALIAN ELIZABETHAN THEATRE TRUST Arts organization established in 1954 (commemorating the first visit of Queen Elizabeth II to Australia) to develop the performing arts in Australia. A non-profit-making public company, it was founded by public subscription and a Commonwealth Government grant. The Trust, acting in an entrepreneurial role, presented many pro-

*Members of the original Light Horse of the first **Australian Imperial Force** before leaving Australia at the beginning of the First World War. (Australian War Memorial)*

ductions, including drama, opera, ballet, musicals, and concerts, with performers from Australia and other countries. It established the Elizabethan Trust Opera Company (1956), from which the Australian Opera (1970) developed; the Marionette Theatre of Australia; the Elizabethan Sydney Orchestra (1967) and the Elizabethan Melbourne Orchestra (1969), to accompany the Australian Opera and the Australian Ballet; and the Theatre of the Deaf (1979). It also helped to establish the Australian Ballet (1962) and many theatre companies in various parts of Australia. The Trust's other main role was to provide administrative services. The Australian Council for the Arts (founded 1968, succeeded in 1973 by the Australia Council) took over some of the Trust's functions. The Trust went into liquidation in March 1991.

AUSTRALIA-NEW ZEALAND AGREEMENT *Anzac Pact

AUSTRALIAN FLYING CORPS *Royal Australian Airforce

AUSTRALIAN IMPERIAL FORCE (AIF) The name given to two military forces whose members were all volunteers. The first AIF was the main Australian expeditionary force during the First World War. *Bridges, who organized it in 1914, also named it. The second AIF was the comparable body during the Second World War.

AUSTRALIAN INDUSTRIAL RELATIONS COMMISSION *Commonwealth Conciliation and Arbitration Commission

AUSTRALIAN LABOR PARTY This democratic socialist party is Australia's oldest political party. The trade union movement organized colonial political parties from 1891 onwards. The failure of the *Maritime Strike and the *Shearers' Strikes acted as a catalyst in the formation of such parties. The Trades and Labor Council formed Labor Electoral Leagues in New South Wales, the Trades Hall Council formed the Progressive Political League in Victoria, and similar bodies with various names were formed in other colonies. The Labor Party has governed federally a number of times (1904, 1908–9, 1910–13, 1914–16, 1929–32, 1941–9, 1972–5, 1983–), and for various periods in the states. During the First World War it split, basically over the issue of *conscription, when *Hughes and his supporters left and helped to form the *Nationalist Party. During the *Great Depression it split again, basically over the issue of fiscal measures to combat the Depression; *Lyons and his supporters left and helped to form the *United Australia Party. In the mid-1950s it split for a third time, basically over the issue of communism; expelled members formed the Australian Labor Party (Anti-Communist), which later became the *Democratic Labor Party. Federal parliamentary leaders have been *Watson (1901–7), *Fisher (1907–15), *Hughes (1915–16), Frank Gwynne Tudor (1917–22), Matthew Charlton (1922–8), *Scullin (1928–35), *Curtin (1935–45), *Forde (1945), *Chifley (1945–51), *Evatt (1951–60), *Calwell (1960–7), *Whitlam (1967–78), William George Hayden (1978–83), *Hawke (1983–91), and *Keating (1991–).

AUSTRALIAN NATIONAL FLAG A flag that has been in use for most of the twentieth century. In 1901 the Commonwealth Government and a magazine entitled

AUSTRALIAN NATIONAL FLAG

Australian Labor Party poster for the 1928 federal elections.

The *Review of Reviews* held a competition to choose a design for a national flag. The Havelock Tobacco Company also contributed to the prize money, which totalled £200. The competition attracted almost 33 000 designs from

Australia and elsewhere. Five different competitors submitted the same winning design, which formed the basis for the present flag. The design was given royal approval in 1903, was changed slightly in 1909, and was officially adopted in 1953. The *Flags Act 1953* describes it in the following way:

> The Australian National Flag is the British Blue Ensign, consisting of a blue flag with the Union Jack occupying the upper quarter next the staff, differenced by a large white star (representing the six States of Australia and the Territories of the Commonwealth) in the centre of the lower quarter next the staff and pointing to the centre of the St George's Cross in the Union Jack and five white stars, representing the Southern Cross, in the fly, or half of the flag farther from the staff.

AUSTRALIAN NATIVES' ASSOCIATION (ANA) A friendly society ('for mutual insurance against distress in sickness or old age'). The Victorian Natives' Society (later Association) was founded in Melbourne in 1871, and restricted its membership to Victorian-born men over sixteen. In 1872 it changed its name to Australian Natives' Association and admitted Australian-born men over sixteen. Women were not admitted as members until 1964. The Association has been active in public debate, advocating causes that it has seen as being patriotic, notably federation, female *suffrage, defence, restriction of *immigration, and conservation. Numerous prominent Australians have been members.

AUSTRALIAN NUCLEAR SCIENCE AND TECHNOLOGY ORGANIZATION (ANSTO) *Australian Atomic Energy Commission

AUSTRALIAN OPERA *Australian Elizabethan Theatre Trust

AUSTRALIAN PATRIOTIC ASSOCIATION An *emancipist organization formed in New South Wales in 1835. Its leading members included Sir John Jamison (president), *Wentworth (vice-president), and Dr William Bland, and its London agent was Charles Buller. Its main purpose was to campaign for representative government, which was achieved, to some extent, with the passing of *An Act for the Government of New South Wales and Van Diemen's Land 1842*, which provided for a New South Wales Legislative Council consisting of thirty-six members, of whom twenty-four were to be elected and twelve nominated.

AUSTRALIAN RULES FOOTBALL *Football

AUSTRALIAN SECURITY INTELLIGENCE ORGANIZATION (ASIO) One of Australia's main security organizations. It

*Menu for an **Australian Natives' Association** luncheon held on Australia Day 1935. (Australian Natives' Association)*

*Kaz Cooke takes a look at ASIO (the **Australian Security Intelligence Organization**) in her 1989 cartoon* PSST! What's the Capital of Argentina?. *(La Trobe Collection, State Library of Victoria)*

*The **Australian War Memorial**, Canberra. (Australian War Memorial)*

was established by the Chifley Labor Government in 1949.

AUSTRALIAN WAR MEMORIAL (AWM) *Bean was largely responsible for the establishment of this statutory authority, the main functions of which are to commemorate the Australians who served in the wars in which Australia has participated, to collect records and promote research, and to display relics and art. Its building was opened in 1941, and has become one of Canberra's major tourist attractions.

AUSTRALIAN WORKERS UNION (AWU) William Guthrie *Spence was mainly responsible for founding this union in 1894, when a number of bush unions in New South Wales, Victoria, South Australia, and Queensland merged. They included the Amalgamated Shearers' Union and the General Labourer's Union. The estimated original membership was 30 000 men. The AWU, Australia's largest trade union for many years, did not affiliate with the *Australian Council of Trade Unions until 1967. In 1992 its membership, drawn from a diversity of industries, numbered about 125 000.

'AUSTRALIA WILL BE THERE' A popular Australian song during the First World War, described by its publishers as 'Australia's Battle Song', the 'Australian War Song', and 'Australia's Marseillaise'. Its title became a patriotic slogan.

AUSTRALIND A proposed model settlement, to be based upon *Wakefield's ideas, on the Leschenault Estuary (near present-day Bunbury) in Western Australia. The Western Australian Company, which was constituted in 1840, planned to sell the land from England. The settlement collapsed within three or four years of the arrival of its first settlers in 1841. Suggested reasons for its failure include lack of labour, the settlers' lack of farming experience, financial mismanagement on the part of the Company, and the unsuitability of the area for small farms. A small town, also named Australind, now exists on the site.

AWM *Australian War Memorial

AWU *Australian Workers Union

B

BABY BONUS Popular name given to a maternity allowance introduced by the Commonwealth Parliament in 1912. Most Australian mothers (some groups, such as *Aborigines and Asians, were excluded) were paid a lump sum, originally £5, on the birth of each child. There was no means test. During the *Great Depression the amount was reduced and a means test was introduced. The Baby Bonus was abolished in 1978.

BALLARAT REFORM LEAGUE *Eureka Rebellion

BANK OF NEW SOUTH WALES When the bank—the first established in Australia—opened for business in 1817 in premises in Macquarie Place, Sydney, the colony of New South Wales lacked a stable local currency. The bank was to have a nominal capital of £20 000; £1425 was subscribed when it opened. *Macquarie, who strongly supported the establishment of such a bank, granted it a charter of incorporation. The British Government declared the charter invalid, but an amended charter was issued in 1823. The Bank of New South Wales was reconstituted and incorporated under an Act of the Legislative Council in 1850, and eventually branches were established throughout Australia and elsewhere. The bank was one of the few to remain open throughout the bank crashes of 1893, and during the *Great Depression it led the way in devaluing the Australian pound in 1931. It took over various smaller banks, and finally merged with the Commercial Bank of Australia in 1981; in the following year it adopted the name Westpac Banking Corporation.

BANK NATIONALIZATION A controversial political issue in Australia between 1945 and 1949. The *Banking Act* 1945 gave the Commonwealth Government increased control over the banking system, but in 1947 the *High Court found some parts of the legislation to be invalid. The *Chifley Labor Government then passed the *Banking Act* 1947, which provided for the nationalization of the banking system. This move led to controversy and opposition, and in 1948 the High Court found some of this legislation, too, to be invalid. The Privy Council upheld the High Court's ruling in 1949. The bank nationalization issue has been seen as a major factor in the defeat of the Chifley Government in 1949.

BANKS, SIR JOSEPH (1743–1820), naturalist, was born in England and educated at Harrow, Eton, and Oxford. He inherited considerable wealth; studied natural science for much of his life, taking part in many expeditions and amassing large collections of animals, plants, rocks and other objects; and acted as a patron to many collectors and explorers. Banks, who has been described as the 'Father of Australia', took part in James *Cook's expedition in the *Endeavour* (1768–71), recommended *Botany Bay as the site for a proposed convict settlement to a British House of Commons committee in 1779, and later gave advice about the new settlement. He corresponded with *Phillip, *Hunter, *King, *Bligh, *Macquarie, and early colonists, and paid the expenses of collectors and explorers such as Alan Cunningham. He also pursued similar interests in other parts of the world. He was President of the Royal Society from 1778 until his death at Isleworth, England.

BARAK, WILLIAM (1824–1903), Abori-

Established in 1817, the Bank of New South Wales was the first to open in the gold-fields. (Westpac Banking Corporation)

*Sir Redmond **Barry** presides over his most famous trial, that of Ned Kelly, in this illustration from the Australasian Sketcher, 20 November 1880. Barry died on 23 November 1880, less than two weeks after Kelly was hanged on 11 November. (National Library of Australia)*

ginal spokesman and artist, was a member of the Wurundjeri clan of the Woiwurung people, who were dispossessed of their land along the Yarra and Plenty Rivers when *Melbourne was settled by Europeans. Barak, who was said to have seen William *Buckley and John Batman during his childhood, was educated at a mission school near Geelong between 1837 and 1839, before becoming a member of the *native police. After moving to Acheron in 1859, he settled permanently at the Coranderrk Aboriginal Reserve, near Healesville, in 1863, working on its farm. 'King Billy, last chief of the Yarra Yarra tribe', as he was called, became a respected and influential spokesman on Aboriginal affairs from the 1870s onwards. Since described as 'an accomplished painter in ochre and charcoal' and 'the first Aboriginal artist of renown', Barak recorded aspects of Aboriginal life in drawings during the 1880s and 1890s. He died at Coranderrk.

BARCALDINE A town in central western Queensland. During the *Shearers' Strikes of 1891, shearers established a

*Edmund **Barton**, Australia's first prime minister (1901–3), was the acknowledged leader of the federation movement in New South Wales during the 1880s and 1890s. This photograph of him was used in the album of delegates to the third session of the Federal Convention, held in Melbourne in 1898, of which he was the leader. (Australian Natives' Association)*

number of camps, the best-known of which was at Barcaldine, where about one thousand shearers, said to be armed and advocating incendiarism, were camped. They flew the *Eureka flag. A number of union leaders were sentenced to terms of imprisonment following their arrests at Barcaldine.

BARRY, SIR REDMOND (1813–80), lawyer, was born in Ireland, educated there and in England, and was appointed to the Irish Bar in 1838, before migrating to Australia in 1839. He practised as a barrister in Melbourne, and became Commissioner of the Court of Requests in 1843, Victoria's first solicitor-general in 1851, and a judge of the Victorian Supreme Court in 1852. Well-known trials over which he presided include those of Raffaello Carboni and others involved in the *Eureka Rebellion, and that of Ned *Kelly. Barry's many contributions to Melbourne's development include helping to found the Melbourne *Mechanics' Institute, the University of Melbourne (of which he was the first chancellor, from 1853 until his death), the Public Library and the Art Gallery. He died in Melbourne.

BARTON, SIR EDMUND (1849–1920), politician, was born in Sydney. His father was a financial agent and sharebroker; his mother ran a girls school during the 1860s. Barton was educated at Fort Street Model School, Sydney Grammar School, and the University of Sydney. He was admitted to the Bar in 1871 and was appointed Queen's Counsel in 1889. He entered state politics in 1879 and was a Member of the Legislative Assembly (1879–87, 1891–4, 1899–1900) and the Legislative Council (1887–91, 1897–8). He held various portfolios. Barton became the acknowledged leader of the *federation movement in New South Wales. During the late 1880s and 1890s he played a prominent role in conferences, set up branches of the Australasian Federation League, and campaigned for the referendums. In 1900 he led the Australian delegation to London to explain the Commonwealth of Australia Constitution Bill. Barton was Australia's first Prime Minister and Minister for External Affairs (1901–3), leading a Protectionist Ministry. The first Commonwealth Parliament was noted for legislation relating to tariffs, *immigration, and defence. Barton was the senior puisne justice on the new *High Court of Australia from 1903 until his death. He supported controversial judgements maintaining states' rights before 1914.

BARWICK, SIR GARFIELD EDWARD JOHN (1903–), politician and lawyer, was born at Stanmore, New South Wales. His father was a compositor. Barwick was educated at Fort Street High School and the University of Sydney. He was admitted to the Bar in 1927, and was appointed King's Counsel in 1941. He was the leading counsel for the banks in the *bank nationalization controversy. Barwick also entered federal politics, and was a Liberal Member of the House of Representatives from 1958 to 1964. He was Attorney-General (1958–64) and Minister for External Affairs (1961–4). Barwick was the Chief Justice of the *High Court from 1964 to 1981; his judgements tended to favour a strict interpretation of legislation controlling industry and imposing taxation. He gave the Governor-General an opinion during the political crisis of 1975, an act seen by some people as controversial.

BASIC WAGE A principle introduced by the *Harvester Judgement, made by *Higgins, the President of the *Commonwealth Conciliation and Arbitration Court, in 1907. Higgins defined it as being sufficient to cover 'the normal needs of the average [male] employee regarded as a human being living in a civilised community'. It was supposed to be for a family of about five, until the introduction of child endowment during the Second World War. Margins (as they became called) for skills were added to the basic wage, and cost-of-living adjustments were made automatically after 1913. A royal commission recommendation in 1920 that the basic wage should be higher was not implemented. During the *Great Depression, the real basic wage was reduced by 10 per cent in 1931, much of the reduction being restored in 1934. In 1937 the basic wage was divided into a needs wage and a prosperity loading. A basic wage for women, comprising 75 per cent of the male wage, was introduced in 1950. Automatic cost-of-living adjustments were dropped in 1953; it was no longer considered to be a needs wage, being above such a minimum. By then the real basic wage was about 50 per cent higher than in 1911, and the working week had decreased by about 20 per cent. The Commonwealth Conciliation and Abitration Commission replaced the idea of the basic wage plus margins with that of the total wage in 1967. A minimum wage for men was then prescribed. The wage-fixing bodies in the various states also adopted the principle of the basic wage, and later that of the total wage, although some variations occurred.

BASS, GEORGE (1711–1803?), naval surgeon and explorer, was born in Lincolnshire, England, and arrived in New South Wales in the *Reliance* in 1795. In that year Bass, *Flinders and William Martin explored parts of *Botany Bay in the *Tom Thumb*. In 1796 the three sailed south to present-day Lake Illawarra, in a second *Tom Thumb*. Bass unsuccessfully attempted to cross the Blue Mountains, and made various other explorations on land. After sailing in the *Reliance* to South Africa, Bass searched successfully for coal at Coalcliff. Late in 1797 he travelled about 1930 kilometres south in a whaleboat, discovered the Shoalhaven River, Twofold Bay, Wilson's Promontory, and Western Port, and deduced the existence of a strait between the mainland and *Van Diemen's Land. Bass and Flinders sailed south in the *Norfolk* in 1798, confirmed the existence of the strait (now called Bass Strait) and circumnavigated Van Diemen's Land, before returning to Port Jackson in 1799. Bass also undertook scientific investigation, and was elected a member of the Linnean Society in London in 1799. In that year he left New South Wales, on medical grounds, and returned to England, where he was given leave. He and Charles Bishop became partners in several commercial ventures. They bought the *Venus* and sailed for New South Wales with a cargo of goods to sell, but the venture failed. They then bought pork and salt in the South Pacific islands for *King; this venture was successful. In 1803 Bass sailed in the *Venus* with contraband to sell in South America; nothing more was heard of him or the *Venus*.

BATAVIA A Dutch East India Company ship. In October 1628 it left Holland for Batavia (present-day Jakarta), under Francisco Pelsaert's command, carrying more than 300 soldiers, crew and passengers. In June 1629 it struck a reef off the Houtman Abrolhos (some 75 kilometres from present-day Geraldton, Western Australia). Some of those on

The surviving stern section of the **Batavia**, *excavated, recovered, preserved, and reconstructed by the Western Australian Museum's Department of Maritime Archaeology, and displayed at the Western Australian Maritime Museum, Fremantle. (Patrick Baker, Western Australian Maritime Museum)*

board drowned; others reached nearby islands. Pelsaert, with a number of others, sailed an open boat to Batavia, about 3220 kilometres away. Meanwhile the ship's supercargo, Jerome Cornelius, led a group of mutineers who planned to take silver from the cargo, murder some of the survivors, and seize the rescue ship. They killed more than 125 men, women and children. Webbye Hayes, a soldier, led a group calling itself the 'Defenders', which captured Cornelius and killed some of the mutineers. When Pelsaert returned in the *Saerdam*, the mutineers were tried and tortured. Seven, including Cornelius, were hanged, and two others were marooned on the mainland. Pelsaert and the survivors sailed for Batavia. Some of the mutineers were punished during the voyage, and some were retried in Batavia, where there were further hangings in 1630. The wreck of the *Batavia* was discovered in 1961. Part of it has since been salvaged and reconstructed.

BATES (née O'Dwyer), **DAISY MAY** (1863–1951), anthropologist and welfare worker, was born in Tipperary, Ireland, but later lived in London. In March 1884, soon after migrating to Australia, she appears to have married the man who was to become famous as 'Breaker' *Morant, in the Queensland town of Charters Towers. They soon separated and she took up a position as a governess at Berry, in New South Wales, later that year. She married a cattleman named Jack Bates in February 1885 and had a son the following year. Daisy Bates returned to England alone in 1894, and worked as a journalist in London for some years. After her return to Australia in 1899, she spent most of her life living with *Aborigines in Western Australia and South Australia, doing research and welfare work. Daisy Bates, who died in Adelaide, wrote some 270 articles about Aborigines and an autobiographical work entitled *The Passing of the Aborigines* (1938). Her papers are held in the National Library of Australia. Margaret *Sutherland's opera *The Young Kabbarli* (1965) is based upon an episode in Daisy Bates's life.

BATMAN, JOHN *Port Phillip Association

BATTLE OF BRISBANE *Brisbane, Battle of

BATTLE OF PINJARRA *Pinjarra, Battle of

'BATTLE OF THE PLANS' *Great Depression

BAUDIN, THOMAS NICHOLAS (1754–1803), French navigator and naturalist, joined the French navy in 1774. He took part in several scientific expeditions before leading a cartographic and scientific expedition, generally considered to have been fairly unsuccessful, to *New Holland. His ships, the *Geographe* and *Naturaliste*, arrived on the west coast, via Mauritius, in 1801. Nicholas Baudin surveyed from Cape Leeuwin to Cape Levêque; Emmanuel Hamelin, commanding the *Naturaliste*, surveyed from Cape Leeuwin to Shark Bay. After visiting Timor, they sailed to *Van Diemen's

Daisy **Bates**, in her typical Edwardian dress, photographed with a group of Aborigines at Ooldea, South Australia, by A. G. Bolam in 1920. (La Trobe Collection, State Library of Victoria)

Land, arriving there early in 1802, and surveyed the D'Entrecasteaux Channel. Baudin surveyed the south coast of New Holland, westwards from Wilson's Promontory, and met *Flinders at Encounter Bay. He believed that he was the first European to discover this coast, which he named Terre Napoléon, but in fact was the first to discover only the section from Cape Banks to Encounter Bay. After sailing south around Van Diemen's Land he arrived in Sydney in mid-1802. Hamelin surveyed Port Dalrymple and Western Port. Baudin sent the *Naturaliste* back to France from Sydney, and continued the expedition, having bought the *Casuarina*, which he placed under Louis de Freycinet's command, from *King. The expedition proceeded to *King Island (where the British made it clear that Van Diemen's Land was a British possession), then surveyed *Kangaroo Island, the west coast of New Holland, and part of the north-west coast. In 1803 the two vessels reached Mauritius, where Baudin died shortly afterwards. The *Casuarina* was left there, but the *Geographe* arrived back in France in 1804. Baudin's expedition exacerbated British fears that the French might settle in New Holland or Van Diemen's Land.

BAYNTON (née Lawrence), BARBARA JANE (JANET AINSLEIGH) (1857–1929), writer, was born to Irish parents at Scone, New South Wales, and was educated at home. She worked as a governess before the first of her three marriages. Barbara Baynton (who is known by her second husband's surname) alternated between residence in Australia and England. She died in Melbourne. She is best known for *Bush Studies* (1902), a collection of six stories that present an unromantic view of Australian outback life.

BEAN, CHARLES EDWIN WOODROW (1879–1968), journalist and historian, was born at Bathurst, New South Wales, where his father was the headmaster of All Saints' College. Bean was educated in Australia and England, and graduated from Oxford before being called to the Bar of the Inner Temple, London, in 1903. He taught at Brentwood School and as a travelling tutor, before returning to Australia where he was admitted to the Bar in New South Wales in 1904. He again taught, this time at Sydney Grammar School, before going on circuit as a judge's associate in New South Wales from 1905 to 1907. Bean then turned to journalism; he became a reporter on the *Sydney Morning Herald* in 1908, and represented that newspaper in London from 1910 to 1912. Bean was Official War Correspondent with the *Australian Imperial Force during the First World War; he went to Egypt with the first contingent, landed on 25 April 1915 at Gallipoli, where he was later wounded, and continued on to France, where he worked from 1916 to 1918. He then became Official War Historian, a position that occupied him for many years. Bean played an important role in the creation of the *Australian War Memorial, and became chairman of its board in 1952. He founded the Parks and Playgrounds Movement of New South Wales. During the Second World War, Bean worked for the Department of Information for some time. He also chaired the Commonwealth Archives Committee in 1942. He chaired the promotion appeals board for the *Australian Broadcasting Commission from 1947 until 1958. He died in Sydney. Bean has been credited with being one of the main creators of the *Anzac legend, and is best known for his work on the publication of *The Official His-*

tory of Australia in the War of 1914–1918 (1921–42), of which he wrote six volumes and edited the remaining six. He wrote a number of other books about the First World War and other subjects.

BENNELONG (1764?–1813), an *Aborigine, was captured on Governor *Phillip's orders and brought to Sydney Cove for observation in 1789. He escaped in 1790, but later returned to the settlement. A hut was built for him at Sydney Cove in 1791. Its site is now known as Bennelong Point. Phillip took Bennelong to England in 1792, where he was presented to King George III. Bennelong returned to Sydney in 1795. He died at Kissing Point.

BHP *Broken Hill Proprietary Company Limited

BIG BROTHER MOVEMENT A migration assistance organization founded in Australia in 1925 by Richard Linton and others. Its purpose was to assist boys and young men wishing to work in rural occupations (later expanded to cover other occupations) in Australia. The Big Brother Movement sponsored approximately 12 500 boys and young men, almost all from Britain, between 1925 and the end of 1982. Using funds which were mainly privately raised, the organization provided accommodation, food, welfare services, and assistance to find employment, but not fares. Members were known as Big Brothers,

*The first group of British boys and young men to be settled in Australia by the **Big Brother Movement** shortly before sailing on the* Jervis Bay *in November 1925. (Big Brother Movement)*

migrants as Little Brothers. Since 1982 it has provided awards and scholarships for young people.

BIGGE, JOHN THOMAS (1780–1843), judge and royal commissioner, was born in Northumberland, England, and educated at Newcastle Grammar School, Westminster School and Oxford. He pursued a legal career, and was Chief Justice of Trinidad from 1813 until his appointment in 1819 to inquire into the colony of *New South Wales. He visited New South Wales and *Van Diemen's Land for about eighteen months between 1819 and 1821, and produced three official reports in 1822 and 1823. These are usually referred to as the *Bigge reports. Bigge then helped to conduct a similar inquiry into Cape Colony, Ceylon and Mauritius (1823–9). His health was very poor when he returned to England in 1829. He died in London.

BIGGE REPORTS Popular name given to *Bigge's findings following his inquiry, held between 1819 and 1821, into the colony of *New South Wales. Criticisms have been made of the way in which the inquiry was conducted. The three reports were *The State of the Colony of New South Wales* (1822), *The Judicial Establishment of New South Wales and of Van Diemen's Land* (1823), and *The State of Agriculture and Trade in the Colony of New South Wales* (1823). Bigge was critical of *Macquarie's treatment of convicts and *emancipists, which Bigge saw as being excessively lenient, and his expensive programme of public works. The reports led to the establishment of a nominated legislative council in New South Wales, the separation of *Van Diemen's Land from New South Wales, a number of legal reforms, changes in the reception of convicts, and the encouragement, by a variety of means, of the pastoral industry.

BJELKE-PETERSEN, SIR JOHANNES (1911–), politician, was born in New Zealand. He farmed in Queensland before becoming a Member of the Queensland Legislative Assembly in 1947, representing the Country Party (later renamed the *National Party). He was Premier from 1968 to 1987, leading a Country–Liberal Coalition Government from 1968 to 1983, then a National one.

BLACKBIRDING A practice that involved trafficking in Pacific Islanders. During the nineteenth century employers in Australia, Fiji and Samoa recruited Pacific Islanders (then more commonly known as Kanakas, from the Hawaiian word for 'man', South Sea Islanders or Polynesians). An early attempt in 1847 to use Pacific Islanders on New South Wales sheep stations was short-lived. Some 57 000 Pacific Islanders were brought to Queensland and northern New South Wales between 1863 and 1904, mainly from the New Hebrides (present-day Vanuatu) and the Solomon Islands. Most worked on sugar plantations, although a few worked on stations and cotton plantations. Their wages were low. In theory, they voluntarily entered into contracts to work in Australia for fixed terms. In practice, they were subjected to abuses that included kidnapping, slavery, and murder. From the 1860s onwards the Queensland Government passed various Acts relating to the recruitment and employment of Pacific Islanders. The Colonial Office in London also made some attempts to control the situation, insisting that government agents be included on re-

This grim illustration, entitled Blackbird Taming, *published in the* Illustrated Melbourne Post, *1872, gives some indication of the kinds of abuses inflicted upon Pacific Islanders as part of the* **blackbirding** *trade during the second half of the nineteenth century. (National Library of Australia)*

cruiting ships in 1871. The British Government passed the *Pacific Islanders Protection Act* in 1872 and strengthened it in 1875. A Queensland Government royal commission, held in 1884, led to a ban on the introduction of Pacific Islanders after the end of 1890, but the ban was lifted in 1892. The Commonwealth Government legislated to ban the introduction of Pacific Islanders from 1904 onwards, and to deport those who were already in Australia after 1906. Some who fulfilled certain qualifications were allowed to remain.

BLACK FRIDAY Popular name given to Friday, 13 January 1939, a day on which disastrous bushfires in Victoria reached a climax. There had been bushfires in December 1938 and the early days of January, and on several days, including Black Friday, high temperatures and strong winds prevailed. Seventy-one people died in the bushfires. A number of towns, including Noojee, were destroyed, much stock died, and a great deal of property was lost, including many timber mills. Bushfires occurred elsewhere in Australia during the same period.

BLACK LINE A cordon of about two thousand White soldiers and civilians, formed in 1830 when Governor *Arthur decided to capture all of the *Aborigines living in the eastern part of *Van Diemen's Land, or to drive them on to Tasman's Peninsula, in the south-east of the island, where they were to be confined. The Black Line operated for about seven weeks. Only two Aborigines were captured.

BLACK POLICE *Native police

BLACK THURSDAY Popular name given to Thursday, 6 February 1851, a day on which very serious bushfires occurred in the colony of Victoria. There had been other bushfires in the preceding

*Some of the horrors of **Black Thursday**, 6 February 1861, are captured in William Strutt's painting of that name (1862–4). (La Trobe Collection, State Library of Victoria)*

weeks, but those on Black Thursday were the worst. It was a day of extreme heat—the temperature in Melbourne at 11 a.m. was 47°C in the shade—and there were strong northerly winds. Many areas of the colony were burnt. At least ten people, including a woman and her five children, were killed. Stock and property losses were very high.

BLACK WEDNESDAY The name given to Wednesday, 8 January 1878, when the Berry Ministry dismissed several hundred Victorian public servants, including judges and magistrates. The Legislative Assembly and the Legislative Council had reached a deadlock over reform of the latter and *payment of members. The Legislative Council had refused to pass the Appropriation Bill (which included provision for payment of members). Berry claimed that he could not pay those dismissed, because his government had not been voted funds, but he was not unwilling to attack his opponents. Some of those dismissed were later reinstated.

BLAMEY, SIR THOMAS ALBERT (1884–1951), soldier, was born near Wagga Wagga, New South Wales. He was a schoolteacher before joining the *army. He studied and then served in India from 1911 onwards. During the First World War he took part in the *Gallipoli campaign and in fighting in Europe as a member of the *Australian Imperial Force. He held various positions, including that of Chief-of-Staff of the Australian Corps in France in 1918. He retired from the army in 1925, and was Chief Commissioner of Police in Victoria from 1925 to 1936. He later chaired the Manpower Committee. During the Second World War he commanded in turn the 1st Australian Corps, the Australian Imperial Force in the Middle East, and the Anzac Corps. He became Deputy Commander-in-Chief, Middle East, after the fall of Greece. In 1942 he became Commander-in-Chief of the Australian Military Forces and Commander of the Allied Land Forces in the South-West Pacific. He commanded operations in *New Guinea. Blamey retired from the army in late 1945. In 1950 he became the first Australian to be promoted to the rank of fieldmarshal.

BLAXLAND, GREGORY *Blue Mountains, Crossing of the

BLIGH, WILLIAM (1754–1817), naval

officer and *governor, was born at Plymouth in England. He pursued a naval career, sailing with James *Cook's last voyage, surviving the *Bounty mutiny, and commanding ships in the battles of Camperdown (1797) and Copenhagen (1801). He was made a Fellow of the Royal Society in 1801, and succeeded *King as Governor of New South Wales in 1806. His conflict with the *New South Wales Corps culminated in the *Rum Rebellion of 1808. After his return to England in 1810, Bligh was virtually exonerated. He died in England.

BLOODY FRIDAY Popular name given to Friday, 19 June 1931, when policemen and anti-evictionists fought a violent battle at a house in the Sydney suburb of Newtown. During the *Great Depression, many tenants who were unable to pay their rents were evicted. The Communist-led Unemployed Workers' Movement prevented some such evictions. The police, instructed by the *Lang Labor Government, attempted to stop the activities of the Unemployed Workers' Movement, and on the morning of 19 June about forty policemen attempted to enforce an eviction order at 143 Union Street, Newtown. About eighteen people were defending the house. The policemen were armed with guns; the anti-evictionists were armed with makeshift weapons, such as bricks and iron bars. A crowd said to consist of many thousands watched the battle that ensued. Eventually the policemen overpowered the anti-evictionists. Fourteen anti-evictionists and thirteen policemen were injured; some receiving fractured skulls. A further demonstration took place at Railway Square in the evening. More than twenty people were arrested on Bloody Friday.

BLUE MOUNTAINS, CROSSING OF THE A major feat of exploration in 1813. The Blue Mountains, which form part of the Great Dividing Range, prevented westward expansion of White settlement in New South Wales. Numerous attempts to cross them failed. In 1813 Gregory Blaxland, William Lawson and *Wentworth discovered a possible route, but did not actually cross the main range. George William Evans, using the same route, did so later in the year. William Cox, acting on *Macquarie's instructions, supervised the building of the first road across the Blue Mountains in 1814 and 1815. The crossing of the Blue Mountains led to extensive pastoral expansion.

BODYLINE SERIES (1932-3) A controversial Test *cricket series played between England and Australia during the Great Depression. Douglas Jardine captained the English team, Bill Woodfull the Australian. Harold Larwood and Bill Voce used bodyline (or fast leg theory as the English called it) bowling against the Australians. It consisted of fast, short-pitched deliveries, aimed (according to the Australians) at the batsman's body rather than the wicket, with the fielders placed in a leg-side position. The batsman could defend his body and risk being caught, or risk being hit by the ball. The Australians saw it as an attempt to curb *Bradman's high scoring, all other methods having failed. Press and public protest became vigorous during the third Test, in Adelaide, when Woodfull and W. A. Oldfield were injured. After being injured, Woodfull made his now-famous comment in the dressing-room to Pelham Warner, the English manager, that one team was playing cricket and the other was not. During the Test, the

Mick Armstrong's cartoon *Now bring out your Larwood*, referring to the English bowler Harold Larwood, was published in the Melbourne *Herald*, 25 November 1932, after the visiting English team first used bodyline bowling in a preliminary match before the **bodyline series**. (La Trobe Collection, State Library of Victoria)

A group of South Australian nurses shortly before their departure from Australia in 1900 for the Boer War. (Australian Archives, South Australia)

Australian Cricket Board of Control cabled the Marylebone Cricket Club, in London, claiming that bodyline bowling was unsportsmanlike and that 'Unless stopped at once it is likely to upset the friendly relations existing between Australia and England'. The Marylebone Cricket Club rejected Australia's claims and stated that they would agree to cancellation of the rest of the series if Australia wished. The series continued. England won the Ashes, having lost only the second Test in Melbourne. The Marylebone Cricket Club altered its rules in 1935, banning such bowling.

BOER WAR (1899–1902) The second of two wars between the British and the Boers in South Africa. The Australian colonies, as parts of the British Empire, became involved. Enthusiasm was keen, even before the war began; Queensland offered troops, Victoria and New South Wales followed suit, and the military commandants of the six colonies conferred in Melbourne and agreed upon a scheme to send men if necessary. In total, about 16 500 men, of whom about 6800 were Bushmen (mostly untrained volunteers, paid either by public subscription or, later, by the British Government), left Australia for South Africa. After federation in 1901, the Commonwealth took responsibility for the defence forces; accordingly, those fighting in South Africa became known as the Australian Commonwealth Horse. The defence of the Elands River at Brakfontein was the most significant fighting in which the Australians were involved. A small number of Australian nurses also served in South Africa. The war ended with the signing by the British and the Boers of the Treaty of Vereeniging at Pretoria on 31 May 1902. Five hundred and eighty-eight Australians had been killed or had died from disease. The execution of 'Breaker' *Morant and Peter Joseph Handcock during the war inspired controversy.

'BOLDREWOOD, ROLF' *Browne, Thomas Alexander

BOLTE, SIR HENRY EDWARD (1908–90), politician, was born at Ballarat, Victoria, and educated at Skipton State School and Ballarat Grammar School. Bolte, like his father, became a farmer. During the Second World War he served in the *army, from 1940 to 1944. He entered state politics, and was a Liberal Member of the Victorian Legislative Assembly from 1947 to 1972. Bolte was Premier and Treasurer, leading a Liberal (the party in Victoria was known as the Liberal and Country Party for some of these years) Ministry from 1955 to 1972. He also held various other portfolios between 1948 and 1950, and from 1955 onwards. Bolte strongly supported states' rights, and his support for capital punishment aroused controversy. He retired in 1972.

BOROVANSKY, EDOUARD (1902–59), ballet dancer, choreographer, and director, was born in Přerov (in present-day Czechoslovakia), and was educated at the Prague National Theatre Ballet School (and also belonged to its company for some time). He first visited Australia in 1929, with Anna Pavlova's company. Borovansky was a soloist with the Ballet Russe de Colonel de Basil from 1932 to 1939, when he left during an Australian tour. He began a ballet school in Melbourne in 1939, and an affiliated ballet club was formed the next year. His students performed as the Borovansky Australian Ballet in 1940. The Borovansky Ballet, as it became known, continued (with several breaks) until 1961, and then formed the basis for the Australian Ballet, established in 1962. The former's best-known dancers included Kathleen Gorham, Peggy Sager, Paul Hammond, Garth Welch, and Marilyn Jones, and Dame Margot Fonteyn appeared with the company as a guest in 1957. Its repertoire covered classical works and some modern ballets, including Borovansky's *Terra Australis*. Borovansky, who has been described as 'the father of Australian ballet', died in Sydney.

BOTANY BAY An inlet on the eastern coast of Australia, about 8 kilometres south of present-day Sydney, where James *Cook first landed in Australia (or *New Holland, as it was then called) on 29 April 1770. The first White person to discover it, he named it Stingray Bay, and stayed there for a week, during which time *Banks observed many new plants, hence its change of name to Botany Bay. The British Government instructed *Phillip and the *First Fleet to settle there, but on his arrival at Botany Bay in 1788, Phillip decided instead to establish the settlement at Port Jackson. During the *transportation era the name Botany Bay was often used as a synonym for New South Wales.

BOUNTY MUTINY A mutiny that occurred near the Tongan Islands in 1789. HMS *Bounty*, under *Bligh's command, was carrying a cargo of bread-fruit trees from Tahiti to the West Indies when the crew mutinied. Bligh and eighteen

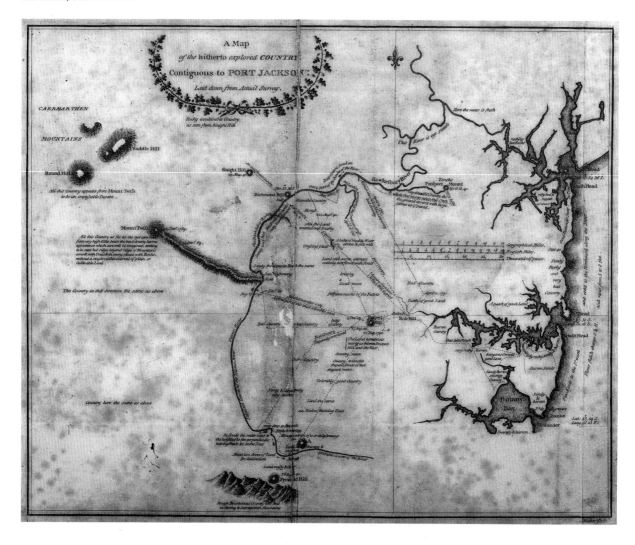

*A 1791 map of Port Jackson, Broken Bay, and **Botany Bay** by J. Walker, based upon the Dawes map. (Mitchell Library, State Library of New South Wales)*

officers were cast off in a small open boat, in which they sailed to Timor, a distance of some 3618 miles, in about six weeks. Bligh was exonerated at a court martial in England. Some of the mutineers surrendered and others were captured in 1791. The frigate *Pandora*, taking them back to England, sank in Torres Strait. The victims included some of the mutineers. Ten of the mutineers were eventually court-martialled in England; four were acquitted, and six were sentenced to death (but only three were hanged). Fletcher Christian and some of the other mutineers, with a number of Tahitians, settled on Pitcairn Island in 1790. Their descendants moved to *Norfolk Island in 1856. Bligh's account of the mutiny, *A Narrative of the Mutiny on Board H.M.S. Bounty*, was published in London in 1791. Marine archaeologists investigated the *Pandora* wreck in the 1980s.

BOURKE, SIR RICHARD (1777–1855), *governor, was born in Dublin, and educated at Westminster School and Oxford, before pursuing a military career. From 1826 to 1828 he acted as Governor of Cape Colony, where he was responsible for ending censorship of the press, tried to reduce border conflict with the Kaffirs, and legislated to improve conditions for the Hottentots. He was appointed Governor of New South Wales in 1830, and arrived in Sydney the following year, where he succeeded in extending financial assistance beyond the Church of England to other denominations, although his attempts to assist non-Anglican schools were defeated. He introduced a 'bounty' system of *immigration. He also introduced licences giving *squatters the right to use Crown lands. His conflict with *exclusives during his term of office reached a climax in an altercation with Campbell Riddell, the Colonial Treasurer, which led to Bourke's resignation in 1837.

BOXER RISING A Chinese rebellion in 1900 against Europeans in China. A contingent of about 460 naval volunteers from Victoria and New South Wales went to China to assist Britain and other European countries in suppressing the rebellion. The men saw some fighting, but were mostly involved in non-combatant duties at Peking and Tientsin. They were withdrawn in 1901. Six Australians died during the Boxer Rising.

BOYD, BENJAMIN (1803?–51), entrepreneur, was born in London, and became a stockbroker. Boyd sailed to Australia in 1842, having been promised some government assistance with his ventures, which were to be funded by the bank he had founded in 1839, the Royal Bank of Australia. He established a Sydney branch of the bank; ran a steamship service between Sydney, Twofold Bay, and Hobart Town; became a *squatter, with properties in the Monaro district and the *Port Phillip District; and established a wool-cleaning establishment and stores at Neutral Bay. In 1847 he brought about two hundred Pacific Islanders to Australia to use as cheap labour, a forerunner of *blackbirding, but the scheme failed within a year. Boyd founded two small towns at Twofold Bay, Boyd Town and East Boyd (which was the site of a whaling station), and represented the Port Phillip District in the Legislative Council of New South Wales from 1844 to 1845. In 1847 he was dismissed as director of the Royal Bank, which went into liquidation in 1849. Boyd sailed his yacht to California in 1849, but failed at the gold-fields, and in 1851 returned to the Pacific Islands, where he was apparently murdered in the Solomon Islands. A search for him in 1854 was unsuccessful.

BOYD FAMILY A well-known Australian artistic and literary family. Its members include ARTHUR MERRIC BOYD (1862–1940), painter, notably of water-colours; ARTHUR MERRIC BLOOMFIELD BOYD (1920–), painter, whose paintings include the *Love, Marriage and Death of a Half-Caste* series, first exhibited in 1958; DAVID FIELDING GOUGH BOYD (1924–), potter and painter; GUY MARTIN À BECKETT BOYD (1923–), sculptor; MARTIN À BECKETT BOYD (who sometimes used the pseudonym 'Martin Mills') (1893–1972), writer, whose novels, some of which are based upon his own family, include *The Montforts* (1928), *Lucinda Brayford* (1946), *The Cardboard Crown* (1952), *A Difficult Young Man* (1955), *Outbreak of Love* (1957), *When Blackbirds Sing* (1962), and *The Tea Time of Love* (1969); ROBIN GERARD PENLEIGH BOYD (1919–71), architect and writer, whose books include *Australia's Home* (1952), *The*

An 1829 lithograph of Sir Richard Bourke, before he became Governor of New South Wales. (National Library of Australia)

Australian Ugliness (1960), and *The Great Australian Dream* (1972); THEODORE PENLEIGH BOYD (1890–1923), painter and etcher; and WILLIAM MERRIC BOYD (1888–1959), potter.

BOYD TOWN *Boyd, Benjamin

'BRADDON'S BLOT' Derogatory term used to describe Section 87 of the *Constitution of the Commonwealth of Australia. It was named after Sir Edward Braddon, Premier of Tasmania from 1894 to 1899, who was responsible for its introduction. Section 87 says:

> *During a period of ten years after the establishment of the Commonwealth and thereafter until the Parliament otherwise provides, of the net revenue of the Commonwealth from duties of customs and of excise not more than one-fourth shall be applied annually by the Commonwealth towards its expenditure. The balance shall, in accordance with this Constitution, be paid to the several States, or applied towards the payment of interest on debts of the several States taken over by the Commonwealth.*

A new financial arrangement, less favourable to the states, was made in 1910; it gave annual grants to the states of twenty-five shillings per head of population. But this was not in the Constitution, and could be repealed by Parliament at any time. It was repealed under the *Financial Agreement.

BRADFIELD, JOHN JOB CREW *Sydney Harbour Bridge

BRADMAN, SIR DONALD GEORGE (1908–), cricketer, was born at Cootamundra, New South Wales, and educated at Bowral Intermediate High School. Bradman played his first interstate match for New South Wales in 1927. He went on to play in many Sheffield Shield and Test matches (including the *bodyline series). He captained the Australian team in Test matches against England from 1936 to 1948. Bradman has often been described as the greatest batsman in the history of *cricket. He scored 8926 runs, averaging 110 runs, in Sheffield Shield cricket; 6996 runs, averaging 99.9 runs, in Test cricket; and 28 067 runs, averaging 95 runs, in all first-class cricket. He held the world's record score in first-class cricket of 452 not out. He chaired the Australian Board of Control for International Cricket from 1960 to 1963, and from 1969 to 1972. Bradman was also a stock and share broker. His publications include *Farewell to Cricket* (1950), *How to Play Cricket* (1955), and *The Art of Cricket* (1958).

BRENNAN, CHRISTOPHER JOHN (1870–1932), poet, was born to Irish parents in Sydney. His father was a brewer, later a publican. Brennan was educated at the Good Samaritan School and Convent, Sydney; St Aloysius's College; St Ignatius's College, Riverview; and the University of Sydney and the University of Berlin. He did not complete his doctorate. Before travelling to Berlin in 1892, Brennan taught at St Patrick's College, Goulburn, and St Ignatius's College, Riverview; after returning, he worked as a cataloguer, and later as an assistant librarian in the Public Library of New South Wales (1895–1909). He pursued an academic career at the University of Sydney: he was a temporary lecturer in languages (1896–7, 1908–9) and in classics (1908) before being appointed to a permanent position in French and German in 1909; he became

Portrait of Sir Edward Braddon, Premier of Tasmania (1894–9), whose name was immortalized in the term 'Braddon's Blot'. (Australian Natives' Association)

Associate Professor of German and Comparative Literature in 1921, but lost his position in 1925 because he had committed adultery. Brennan taught again in schools, at the Marist Brothers' High School, Darlinghurst (1927), and at St Vincent's College, Potts Point (1930–2). He died in Sydney. Brennan is best known for his poetry, which was influenced by the French Symbolists. His publications include *XVIII Poems: Being the First Collection of Verse and*

*Fans reach out to touch Don **Bradman**, Australia's most famous cricketer and a participant in the controversial bodyline series. (The Herald & Weekly Times Ltd)*

Prose (1897), *XXI Poems (1893–1897): Towards the Source* (1897), *Poems* (1914), and *A Chant of Doom and Other Verses* (1918).

'BRENT OF BIN BIN' *Franklin, Stella Maria(n) Sarah Miles

BRIDGES, SIR WILLIAM THROSBY (1861–1915), soldier, was born at Greenock, Scotland. His father was a naval officer. Bridges was educated at the Royal Naval School, New Cross, London, and at the Royal Military College of Canada, but did not complete his course at the latter. He followed his family to New South Wales in 1879, and worked in the civil service there for some years before joining the New South Wales permanent forces in 1885. He served with the British cavalry in the *Boer War for a few months, but was invalided to England with enteric fever. He became the first Chief of the Australian General Staff in 1909. Some months later, he went to England, where he represented Australia on the Imperial General Staff. Bridges established the Royal Military College at Duntroon in 1911, and remained as its Commandant until 1914. As Inspector-General of the Australian Military Forces, he raised the *Australian Imperial Force at the outbreak of the First World War, and was appointed, with the rank of major-general, to command it. His division was the first to land at Anzac Cove on 25 April 1915, during the *Gallipoli campaign. Bridges was wounded in Monash Valley on 15 May 1915 and died three days later, while on the way to hospital in Egypt. He was knighted the day before his death. His remains were buried in the Australian Capital Territory.

BRISBANE, the capital city of *Queensland, lies on both sides of the Brisbane River, about 20 kilometres from the river mouth in *Moreton Bay. A *penal settlement, named after the Governor of New South Wales, was established there in 1824 and operated until 1839. Free settlement was forbidden until 1842, the year of the first Brisbane land

Queen Street, Brisbane, in 1860, only eighteen years after free settlement was allowed. (La Trobe Collection, State Library of Victoria)

sales. Brisbane became a city in 1902. Its population was about 5000 in 1859, 25 916 in 1871, 101 554 in 1891, and 1 240 300 in 1988.

BRISBANE, BATTLE OF Popular name for one of the most serious outbreaks of violence between Australian and American servicemen while the latter were stationed in Australia during the Second World War. Australian civilian and military police and American military police were called to a number of brawls in Brisbane on 26 November 1942. During the fighting, which lasted about three hours, one Australian soldier was killed and a number of Australians and Americans were seriously injured. Despite added police precautions, further fighting broke out the following evening, lasting for about four hours.

BRISBANE LINE Popular name given to a supposed plan to defend only the more populated areas of Australia (the settled areas south-east of Brisbane) against Japanese attack during the Second World War. E. J. Ward, the Minister for Labour and National Service in the *Curtin Labor Government, repeatedly alleged that previous non-Labor governments had made such a plan, and referred to it in Parliament as the Brisbane Line in 1943. The allegations were denied. Curtin stated that the only such plan was one prepared by military leaders, discussed by the Advisory War Council in 1942, and rejected by politicians from both sides. Ward then claimed that an important document was missing from the files. A royal commission was held into this claim, during which time Ward was relieved of his portfolio; it found that there was no missing document.

BRISBANE, SIR THOMAS (MAKDOUGALL from 1826) (1773–1860), soldier, astronomer and *governor, was born near Largs, in Ayrshire, Scotland, and was educated by tutors, at the University of Edinburgh, and at the English Academy, Kensington. He then pursued a military career and fought in the Peninsular War, attaining the rank of general. Brisbane was in command of the Munster district in Ireland when he was appointed Governor of New South Wales in 1820. He relieved *Macquarie the following year. His instructions were based largely upon *Bigge's recommendations. A small nominated Legislative Council was established during his term of office. He faced problems because of the conflict between *emancipists and *exclusives in the colony, and he clashed with Frederick Goulburn, the Colonial Secretary. Brisbane protested to the Colonial Office over Goulburn's conduct, and both Brisbane and Goulburn were recalled in 1824. He left the colony in 1825, and was succeeded by *Darling. Brisbane was a keen astronomer; he built two observatories in Scotland and one in Australia (at *Parramatta).

BROKEN HILL A mining centre situated in the Barrier Range, New South Wales, about 1100 kilometres west of Sydney. *Sturt explored the area and named it after its physical characteristics in 1844. Silver, lead, and zinc were found in the area during the 1870s. Charles Rasp, a boundary rider, found the main lode in 1883; it was to become world-famous. The *Broken Hill Proprietary Company began there. Broken Hill, sometimes known as the 'Silver City', is significant because of the extent of its mineral deposits, and because of the strength of the Barrier Industrial Council, which

Leaflet published in 1920 during a strike in Broken Hill, which has a strong history of trade unionism. (National Library of Australia)

has represented unions there since the Council's foundation in 1923. Mining continues in the area (which is known also for its wool-growing), but has been declining in recent years.

BROKEN HILL PROPRIETARY COMPANY LIMITED (BHP) A major Australian company, BHP was formed in 1885 to mine silver, lead, and zinc deposits at *Broken Hill, where it continued to mine until 1939. It began smelting there in 1886 (continuing to do so until 1898) and at Port Pirie in South Australia from 1889 onwards, and took up an iron ore lease at Iron Knob in 1899. In 1915 BHP opened a steelworks at *Newcastle, and bought collieries around there after the First World War. It also acquired a shipping fleet during the 1920s. In 1935 BHP took over Australia's only other steel maker, Australian Iron and Steel Ltd, which had works at Port Kembla in New South Wales. The Company opened a blast furnace at Whyalla in South Australia in 1941, and also operated a shipping yard there from 1942 to 1977. In 1950 it began mining iron ore for Port Kembla at Yampi Sound in Western Australia, and opened a blast furnace at Kwinana, also in that state, in 1968. In 1964 the 'Big Australian', as BHP has often been called, discovered oil and natural-gas fields in Bass Strait and developed them in conjunction with Esso. It established a steel mill (in partnership, but later bought out its partner, maintaining its monopoly on Australian steel making) in Western Port Bay during the 1970s. In 1976 it joined a consortium to develop oil and natural-gas fields on the North-West Shelf off Western Australia. Australia's largest company for most of the twentieth century, BHP has also developed various other related manufacturing interests.

BROOME Town and port on Roebuck Bay in north-western Western Australia, 2205 kilometres by road from Perth. William *Dampier explored the coast in 1688 and 1699 and Phillip Parker King the bay in 1818. Settlers began pearling in the area during the 1860s and the town, named after Sir Frederick Napier *Broome, the then Western Australian Governor, was established in 1883. Broome became known as the 'Port of Pearls', one of the world's major pearling centres, with some four hundred pearling vessels and more than two thousand workers, many of whom were Japanese. During the 1920s, the focus of the Broome industry switched from pearls to pearl shell for the making of cultured pearls. A cyclone sank twenty pearling vessels and killed one hundred and forty crew off Broome in late March 1935. During the Second World War, approximately seventy people were killed when Japanese aircraft bombed Broome on 3 March 1942. A deep-water jetty was opened in 1956. Broome's tourist industry developed considerably during the 1980s.

BROOME, SIR FREDERICK NAPIER (1842–96), *governor, was born in Canada and educated in England. He farmed in New Zealand for a number of years, and held official positions in Natal and Mauritius before his term of office as the Governor of Western Australia (1883–9). Broome supported the introduction of *responsible government for Western Australia, which was granted in 1890. After leaving Australia he acted as Governor of Barbados for some time, and was Governor of Trinidad from

BROUGHTON, WILLIAM GRANT

Mary Grant **Bruce**'s Billabong books became immensely popular. This frontispiece, 'It seemed to Norah that she pulled Bobs up almost in his stride', by S. E. Campbell, is from an undated reprint of *A Little Bush Maid*, the first book in the series. (Ward Lock Ltd)

1891 to 1896. His publications include *Poems from New Zealand* (1868) and *The Stranger of Seriphos* (1869).

BROUGHTON, WILLIAM GRANT (1788–1853), Anglican clergyman, was born in England, worked as a clerk in the East India Company for some time, and was educated at Cambridge University. He was ordained as a deacon in 1818, and later as a priest. Broughton was appointed Archdeacon of New South

Wales in 1828, arrived in Sydney the following year, and became Bishop of Australia (a new position) in 1836. He opposed *Bourke's religious and educational policies. Broughton built numerous churches and schools, began the building of St Andrew's Cathedral, and helped to found new dioceses in New Zealand and Australia. He became Bishop of Sydney, with authority over the other Australasian bishops, and organized a significant conference of these bishops in 1850. Broughton died in England, while on a trip to discuss church government.

BROWNE (NÉ BROWN), THOMAS ALEXANDER ('Rolf Boldrewood') (1826–95), novelist, was born in London. His father, a shipmaster, brought his family to Australia about 1831. Browne was educated in Sydney at W. T. Cape's school and at Sydney College, and in Melbourne by the Reverend David Boyd. He was a *squatter from 1844 to 1869, at first near Portland, then near Swan Hill, and finally near Narrandera. He then became a government official, holding various positions at Gulgong, Dubbo, Armidale and Albury, in the years 1871–95. After retiring, he lived in Melbourne until his death. Browne wrote under a pseudonym, and is best known for his novel *Robbery Under Arms: A Story of Life and Adventure in the Bush and in the Goldfields of Australia* (1888), much of which was based upon actual events. It was first published as a weekly serial in the *Sydney Mail* (1882–3), and at least one play and several films have been based upon it. Browne wrote many other novels.

BRUCE, MINNIE (MARY) GRANT (1878–1958), writer, was born near Sale, Victoria. Her father, a surveyor, was Irish; her mother was of Welsh descent. Mary Grant Bruce was educated at a girls school in Sale, then moved to Melbourne, where she worked as a secretary and later as a journalist. She spent much of her life in Britain and Europe, and is best known for her many books for children, particularly those in the Billabong series. The fifteen Billabong books, which relate the saga of the Linton family and are set in Victoria, were published between 1910 and 1942.

BRUCE OF MELBOURNE, STANLEY MELBOURNE, 1ST VISCOUNT (1883–1967), politician, was born in Melbourne, and educated at Melbourne Church of England Grammar School and Cambridge University. He managed the London branch of his family's firm, Paterson, Laing & Bruce, and practised at the Bar for some time. During the First World War, Bruce served in the British Army from 1914 to 1917; he fought at Gallipoli, was wounded, and was invalided to England in 1915. After returning to Australia, he was a Member of the House of Representatives (1918–29, 1931–3), representing the Nationalists, then the *United Australia Party. He was Prime Minister and Minister for External Affairs from 1923 to 1929, leading the Nationalist–Country Party coalition, sometimes known as the Bruce–*Page Government. Bruce also held other portfolios. His main policies during this period were summed up in the slogan *'Men, Money, and Markets'. The coalition was defeated, and Bruce lost his own seat, in the 1929 federal elections. Bruce was active in the *League of Nations: he represented Australia at the League (1921, 1932–9), was on the Council (1933–6), and served as President (1936). Bruce was Australia's High Commissioner in London from 1933 to 1945.

Stanley Melbourne Bruce, Prime Minister of Australia (1923–9), shown here looking typically urbane.

*William **Buckley** lived with Aborigines from 1803 until 1835 after escaping from a convict settlement in present-day Victoria. This romantic engraving by W. Macleod,* William Buckley, *depicting him as a 'wild white man', was published in the* Picturesque Atlas of Australasia, *vol. 1 (1886), edited by A. Garran. (La Trobe Collection, State Library of Victoria)*

He played a particularly significant role in representing Australia in London during the Second World War, serving on the British War Cabinet and Pacific War Council (1942–5). After retiring as High Commissioner, he remained in Britain, where he became director of a number of companies. He chaired the World Food Council (1947–51) and the British Finance Corporation for Industry (1947–57). He was the first Chancellor of the Australian National University (1951–61). Bruce died in London.

BUBONIC PLAGUE   An infectious disease caused by bacteria spread by flea-carrying rats. Cases of bubonic plague were found in Australia in January 1900, and about 1200 cases were recorded in the following ten years. Most of these occurred in New South Wales and Queensland, some in Western Australia, and a few in Victoria and South Australia; none were recorded in Tasmania. The worst year was 1900. The bubonic plague scare led to some improvements in hygiene, including the demolition of slums, the disposal of much rubbish, and the destruction of tens of thousands of rats.

BUCKLAND RIVER RIOT   One of a number of anti-Chinese riots that occurred in Australia during the *gold-rushes (the worst being the *Lambing Flat Riots). On 4 July 1857 a mass meeting of White *diggers on the Buckland River goldfield, in north-eastern Victoria, resolved to expel the Chinese from the colony. At the time there were about thirty thousand Chinese people in Victoria, about two thousand of whom were on the Buckland River gold-field. Shortly after the meeting, about one hundred White diggers, allegedly led by Americans, drove the Chinese from that field, destroyed or stole their property, and attacked some of them. Some of the Chinese are believed to have been killed. Three Whites were charged with rioting and one with theft; one was found guilty of 'unlawful assemblage' and the others were acquitted.

BUCKLEY, WILLIAM (1780–1856), so-called 'wild white man', was born in England. In 1802 he was sentenced to *transportation for life for having received stolen goods. In 1803 he and two companions escaped from a short-lived convict settlement near present-day *Sorrento, Victoria. His companions disappeared, but Buckley lived among *Aborigines until 1835, when he surrendered at Indented Head to a party of

*In its early years the **Bulletin** championed the workers' cause, as can be seen in this illustration,* The Workman's Cross: Disunion, *which appeared on 30 May 1891. (La Trobe Collection, State Library of Victoria)*

newly arrived settlers in the *Port Phillip District. Buckley was pardoned, and acted as an interpreter with Aborigines for some time after Melbourne was founded in 1835.

BULLETIN In 1880 *Archibald and John Haynes founded the Sydney *Bulletin*, a weekly journal. It became very popular, especially during the late nineteenth century, and was often fondly referred

'BUNYIP ARISTOCRACY'

Some of the extraordinary amount of equipment taken by the ill-fated **Burke and Wills expedition** *is shown in this lithograph by Massina & Co.,* The Departure of the Burke and Wills Expedition *(1881), which was published in the supplement to the* Illustrated Australian News, *May 1881. (National Library of Australia)*

to as 'the Bushman's Bible'. By the late 1880s about 80 000 copies of each issue were being sold. Its early policies were nationalistic, radical and republican. Archibald edited the *Bulletin* from 1886 to 1902, and A. G. Stephens edited the Red Page, a literary section on the inside front cover, from 1896 to 1906. The *Bulletin* promoted Australian writers, including *Brennan, *Dennis, *Gilmore, *Lawson, 'Breaker' *Morant, A. B. *Paterson, and 'Steele Rudd' (*Davis); and artists, including 'Hop' (Livingston Hopkins), *Lambert, Norman *Lindsay, David Low, and David Souter. During the First World War, the *Bulletin* abandoned its Labor sympathies over the *conscription controversy and became a conservative journal. It was still being published in the 1990s.

'**BUNYIP ARISTOCRACY**' Derogatory term used to describe the hereditary peerage proposed by the Select Committee, chaired by *Wentworth, that drafted the New South Wales constitution. Daniel Deniehy referred to 'Botany Bay aristocrats' and a 'bunyip aristocracy' in a speech on the Bill in 1853. The proposal was dropped. Bunyips are mythical Australian monsters, supposed to live in lakes and swamps.

BURKE AND WILLS EXPEDITION Popular name for the Great Northern Exploration Expedition, a disastrous enterprise that cost seven lives. Robert O'Hara Burke, an Irish-born policeman, led the expedition, organized by the Royal Society of Victoria and funded by that body, the Victorian Government and the public. The expedition's main purpose was to make the first crossing of Australia from south to north, which South Australians also were attempting. The well-equipped expedition left Melbourne in August 1860. William John Wills, an English-born surveyor and meteorologist, became second-in-charge during the expedition. Burke left a party at Menindie, which was to follow on, and established a base at Cooper's Creek in November 1860. William Brahe was left in charge there. Burke, Wills, John King and Charles Gray, with six camels, one horse and three months' provisions, left the Cooper's Creek base in December and reached the edge of the Gulf of Carpentaria in February 1861, thus becoming the first to make the crossing. (*Stuart, the South Australian, became the second in July 1862.) Gray died on the return to Cooper's Creek, but the other three reached the base on 21 April 1861, only to discover that Brahe's party had left for Menindie that day. Burke decided, against Wills's advice, not to follow but to make for Mount Hopeless. Brahe returned to the base fifteen days after leaving it, but saw no signs of Burke, nor of his message, and so left again. Several search parties were sent out, from Melbourne and elsewhere. One, led by A. W. Howitt, found King, who had been living with *Aborigines, in September 1861. Burke and Wills had died from malnutrition and exhaustion at the end of June. The public regarded them as heroes but commissioners appointed by the Victorian Government to inquire into the expedition were critical of Burke and several others.

BURNET, SIR (FRANK) MACFARLANE (1899–1985), medical scientist, was born in Traralgon, Victoria, the son of a bank manager and his wife. He was educated at Traralgon and Terang State Schools, Geelong College, and the University of Melbourne, from which he graduated MB, BS, in 1922. After completing his residency at the Melbourne Hospital in 1922, he worked in pathology at the Walter and Eliza Hall Institute for Medical Research in Melbourne until 1925. Burnet, who had been awarded his MD in 1924, did research into viruses at London's Lister Institute from 1925 to 1927, and completed a PhD at London University. He returned to the Hall Institute as assistant director from 1928 to 1931, during which time he and Dr (later Dame) Jean Macnamara discovered that there were at least two poliomyelitis viruses rather than one as previously

Bushranging Vagaries in New South Wales, *by an unnamed cartoonist, depicting a group of **bushrangers** and their victims, was first published in the* Melbourne Punch, *22 October 1863. (National Library of Australia)*

*Grace **Bussell** became a heroine after saving shipwreck victims in 1875. John Rowell's drawing of her is entitled* She set out for the shore. *(The Victorian Readers)*

believed. After working at the National Institute of Medical Research in Hampstead, England, in 1932 and 1933, Burnet again returned to the Hall Institute as assistant director in 1934. From 1944 to 1965 he was its director and Professor of Experimental Medicine at the University of Melbourne. In 1957 he switched the focus of his research from viruses to immunology. In October 1960 he and Britain's Sir Peter Medawar shared the Nobel Prize for Physiology and Medicine for their work on acquired immunological tolerance, which has contributed to the development of organ transplants. After retiring in 1965 Burnet was guest Professor of Microbiology at the University of Melbourne, and wrote and spoke prolifically on medical and social issues, particularly the relationship between science and society. Burnet, a member of many Australian and international committees, was president of the Australian and New Zealand Association for the Advancement of Science (ANZAAS) in 1957, president of the Australian Academy of Science from 1965 to 1969, and inaugural chairman of the Commonwealth Foundation from 1966 to 1969. His many honours and awards include thirteen honorary doctorates, the Royal Society's Royal Medal (1947) and Copley Medal (1959), the Mueller Medal (1962), the Order of Merit (1958), two imperial knighthoods (1951 and 1969), and one Australian one (1978). He published more than 250 scientific papers and many books, including *Cellular Immunology* (1969), his chief work on immunity; *Changing Patterns: An Atypical Autobiography* (1968); and *Dominant Mammal* (1979), which is about the human condition.

BURRANGONG RIOTS *Lambing Flat Riots

BUSHFIRES *Black Friday; *Black Thursday

'BUSHMAN'S BIBLE' *Bulletin

BUSHRANGERS Law-breakers who lived in the bush. The term came into use in the early nineteenth century. The first bushrangers were escaped *convicts, but the main period of bushranging in the south-eastern colonies (where it was most common) was probably during the 1850s and 1860s. The best-known bushrangers include Martin Cash, *Donahoe, *Gardiner, *Hall, Michael Howe, *Kelly and his gang, Frank McCallum ('Captain Melville'), Dan Morgan, Andrew George Scott ('Captain Moonlite'), and Frederick Ward ('Captain Thunderbolt').

BUSSELL, GRACE (1860–1935), shipwreck heroine, was born near Busselton, Western Australia, the daughter of a *squatter and his wife. On 2 December 1875 the steamer *Georgette*, carrying a crew and forty-eight passengers on a voyage from Adelaide to *Fremantle, was wrecked near Cape Leeuwin. The ship's lifeboat capsized and eight women and children were drowned. Grace and an Aboriginal stockman, Sam Isaacs, rode in and out of the surf on their horses for four hours rescuing the survivors. Grace Bussell, who became known as the 'Grace Darling of the West' after the British lighthouse-keeper's daughter who helped to save nine survivors from a shipwreck in the Farne Islands in 1838, was awarded a silver medal by the Royal Humane Society and a gold watch from the British Government. Her story has become part of Australian folklore.

BUVELOT, ABRAM-LOUIS (1814–88), landscape painter, was born in Switzerland, and studied in Lausanne and Paris. He spent some eighteen years in Brazil, painting and taking photographs, and returned to Switzerland in 1852. Buvelot later visited the East Indies (present-day Indonesia) and India, and migrated to Victoria in 1865. He worked as a photographer in Melbourne for some months before returning to painting and some teaching. Poor health forced him to stop painting in 1884. Louis Buvelot, sometimes known as the 'father of Australian landscape painting', has been seen as an influence on the painters of the *Heidelberg School. Buvelot's paintings include *Summer Evening, near Templestowe* (1866) and *Waterpool at Coleraine* (1869).

C

CALWELL, ARTHUR AUGUSTUS (1896–1973), politician, was born in West Melbourne, and was educated at St Mary's, West Melbourne, and the Christian Brothers' College, North Melbourne. After working as a public servant, he was a Labor Member of the House of Representatives from 1940 until 1972. He was Minister for Information from 1943 until 1949, Australia's first Minister for Immigration from 1945 until 1949 (during which time he implemented an extensive European *immigration programme), and Leader of the Labor Opposition from 1960 to 1967. Calwell was a strong supporter of the *White Australia Policy. He opposed Australia's involvement in the *Vietnam War. His publications include *Be Just and Fear Not* (1972).

CAMPBELL, ROBERT (1769–1846), merchant and pastoralist, was born in Scotland. In 1798 he became a partner in the Calcutta firm of Campbell, Clarke & Co. (Campbell & Co. from 1799 onwards), visited New South Wales and began trading operations. Two years earlier the firm had sent a cargo to the colony aboard the *Sydney Cove*, but the ship was wrecked during the voyage; in 1800 Campbell returned with another cargo. He was largely responsible for founding the colony's sealing industry, and eventually developed trading and importing interests to such an extent that he became known in the colony as the 'father of the mercantile community'. He also became involved in colonial public affairs, and served on the Legislative Council from 1825 until 1843. Campbell died at his property, Duntroon, which later became the site of the Royal Military College.

CANADIAN REBELS Political protesters who took part in the rebellions of 1837 and 1838 in Upper Canada (present-day Ontario) and Lower Canada (present-day Quebec). Some of those involved were executed, others were sent to Bermuda, and 153 were transported to Australia. Ninety-five Upper Canadians (most of whom were American-Canadians) were transported to Van Diemen's Land in 1839 and 1840. Three escaped and returned to Canada; the others began to receive free pardons from 1843 onwards. Eventually almost all returned to Canada. Fifty-eight Lower Canadians (most of whom were French-Canadians) arrived in New South Wales in 1840. They also began to receive free pardons from 1843 onwards, and eventually fifty-four of them returned to Canada.

CANBERRA Capital of the Commonwealth of Australia, situated in the *Australian Capital Territory, the site of which was chosen in 1908 and acquired from New South Wales in 1911. *Griffin was largely responsible for its design. Canberra was named and inaugurated in 1913, and the Commonwealth *Parliament transferred there from Melbourne in 1927. Canberra's population was 1150 in 1921, 56 449 in 1961, and 272 500 in 1988.

CANBERRA PACT *Anzac Pact

CASEY, RICHARD GARDINER, BARON (1890–1976), governor-general, was born in Brisbane, and educated at Melbourne Church of England Grammar School, the University of Melbourne, and Cambridge University. During the First World War he served with the first AIF (1914–19). From 1919 to 1924 Casey worked for mining and engineering firms, and was then Australian External

CASTLE HILL UPRISING

*An unknown artist's view of the **Castle Hill Uprising**, entitled* Convict Uprising at Castle Hill *(1804). (Rex Nan Kivell Collection, National Library of Australia)*

Affairs Liaison Officer in London for some years (1924–7, 1927–31). Casey was a Member of the House of Representatives for the *United Australia Party from 1931 to 1940, holding various portfolios at different times. He resigned to become Minister to the USA for two years from 1940. He was a member of the British War Cabinet and Minister of State resident in the Middle East from 1942 to 1943. From 1944 to 1946 he was Governor of Bengal. He was a Liberal Member of the House of Representatives from 1949 to 1960, again holding various portfolios. Casey was granted a life peerage in 1960, and was Governor-General of Australia from 1965 to 1969. His publications include *An Australian in India* (1946), *Friends and Neighbours* (1954), *Personal Experiences, 1939–46* (1962), and *The Future of the Commonwealth* (1963).

CASTLE HILL UPRISING Convict rebellion in 1804 at Castle Hill, an agricultural settlement about 8 kilometres north of *Parramatta in New South Wales. Several hundred convicts captured Castle Hill on 4 March as part of an unsuccessful plan to capture the Hawkesbury, Parramatta, and Sydney settlements. George Johnston and a detachment of the *New South Wales Corps intercepted the convicts near Toongabbie on 5 March, captured Phillip Cunningham and William Johnston, two of the leaders, and then opened fire, killing nine convicts and wounding others, including Cunningham. *King proclaimed martial law from 5 to 10

March. Cunningham and eight other convicts were hanged, others were flogged, about thirty were sent to *Newcastle, and some suspects were dispersed to *Norfolk Island and elsewhere. The event has been described as an Irish rebellion: Castle Hill had a large Irish population, most of those punished were Irish (Cunningham, for example, had been transported after the United Irishmen's rebellion of 1798), and many who took part apparently were Irish. The convict rebels also adopted the United Irishmen's slogan of 'Death or Liberty'.

CATARAQUI An 800-ton ship which became involved in one of Australia's greatest shipping disasters. The ship was sailing from Liverpool to Melbourne carrying 369 emigrants, including many young unmarried women, 73 children, and 46 crew members. At 4.30 a.m. on 4 August 1845, in monstrous seas, it ran aground on rocks off the southwest coast of *King Island in Bass Strait, as a result of a navigational error. The constable of the straits, David Howie, found the nine survivors—one emigrant and eight crew members—one or two days later. The cutter *Midge* picked them up on 7 September 1845. Partly as a result of the disaster, a number of lighthouses were established in Bass Strait.

CHAFFEY, GEORGE (1848–1932) and WILLIAM BENJAMIN (1856–1926), Canadian-born irrigation pioneers, established successful irrigation settlements in California. In 1885 *Deakin met them there and discussed the possibilities of *irrigation for Victoria. The Chaffey brothers came to Australia in 1886, and with government assistance soon established Australia's first major irrigation settlements. Both were on the Murray River—one at Renmark in South Australia, the other at Mildura in Victoria—and they were to be primarily for fruit-growing. Both settlements suffered from numerous problems, the Chaffey brothers' practices were questioned, and their firm went into liquidation and was taken over by the newly formed Renmark and Mildura Irrigation Trusts. A Victorian Royal Commission largely blamed the Chaffey brothers for Mildura's problems. In 1897 George Chaffey returned to California, where he became involved in further irrigation projects. William Benjamin Chaffey spent the rest of his life in Mildura, where he farmed, founded the Mildura (later Mildara) Winery Pty Ltd, was active in the Mildura and the Australian Dried Fruits Associations, and was involved in local government.

CHANAK CRISIS Diplomatic crisis between Britain and Turkey in 1922, which strained Australia's diplomatic relations with Britain. Australia was a signatory to the Treaty of Sèvres (1920), which supposedly concluded the war (First World War) with Turkey. When the crisis occurred, the British Government's request for dominion troops to reinforce British troops at Chanak was reported to Australian newspapers before the official request reached the Prime Minister, *Hughes. Hughes agreed to send troops if necessary, but complained to the British Prime Minister, Lloyd George, about the lack of consultation. The emergency passed and Australian troops were not sent. Hughes also complained when the British Government did not invite the dominions to participate in the Lausanne Conference, at which the new treaty with Turkey was to be negotiated. In 1924 the Commonwealth Government approved the Treaty of Lausanne (1923), which finally concluded the war with Turkey.

CHARTISTS were members of a British political movement formed in 1837. Their People's Charter of 1838, which gave rise to their name, contained six demands: universal male suffrage, voting by ballot, annual parliaments, payment of members of parliament, equal electoral districts, and abolition of the property qualification for members of parliament. In 1839, 1842, and 1848, Chartists made unsuccessful attempts to present petitions to parliament. The movement then declined, and ended in about 1858. Some one hundred Chartists, including John Frost, were transported to Australia after events in 1839, 1842, and 1848. Their crimes included conspiracy, riot, sedition, and treason. Most were transported as a result of the 'Plug-Plot' Riots of 1842, which were basically strikes against the reduction of wages. Virtually all of the transported Chartists went to Van Diemen's Land; three went to New South Wales. Chartist ideas had some influence in Australia—in organizations such as the Ballarat Reform League, for example, which was involved in the *Eureka Rebellion.

CHAUVEL, SIR HENRY GEORGE (1865–1945), soldier, was born at Tabulam, New South Wales, and educated at the Sydney and Toowoomba Grammar Schools. His father was a grazier. Chauvel served in volunteer forces in New South Wales and Queensland before joining the Queensland permanent forces in 1896. He served in the *Boer War, was Adjutant-General of the Australian forces (1911–14), and held various other military appointments before the First World War. During the war he commanded the 1st Light Horse Brigade in 1914 and 1915 in Egypt and Gallipoli; the 1st Division for several months in 1915 and 1916 in Gallipoli and Egypt; the Australian and New Zealand Mounted Division in 1916 and 1917 in the Sinai campaign; the Desert Column in 1917; and the Desert Mounted Corps from 1917 to 1919 in Palestine. Chauvel has often been described as a hero of the Palestine campaign in particular, and was mentioned in dispatches numerous times during the war. His honours and decorations included the Croix de Guerre and the Order of the Nile (which he was awarded twice). After the war he was Inspector-General of the Australian Military Forces from 1919 to 1930, and also Chief of the General Staff from 1923 to 1930, becoming the first Australian general in 1929. During the Second World War, Chauvel was Inspector-in-Chief of the Volunteer Defence.

CHIFLEY, JOSEPH BENEDICT (1885–1951), politician, was born at Bathurst, New South Wales, educated at the local school and the Patrician Brothers' School, Bathurst, and held various jobs, mostly in the New South Wales Railways, before becoming an engine-driver. He was active in union affairs, was dismissed during the *Wartime Transport Strike of 1917, and was demoted on his re-employment. Chifley was a Labor Member of the House of Representatives from 1928 to 1931, and from 1940 to 1951, and held various positions, including those of Treasurer (1941–9), Minister for Post-War Reconstruction (1942–5) and Prime Minister (1945–9). He introduced *uniform taxation in 1942. His government supported a

Joseph Benedict Chifley, Australia's Prime Minister from 1945 to 1949. (National Library of Australia)

policy of full employment, increased social services, encouraged large-scale *immigration, established the *Snowy Mountains Hydro-Electric Scheme and *Trans Australia Airlines, nationalized *Qantas (its attempt to nationalize domestic airlines failed), attempted unsuccessfully to introduce *bank nationalization (which Chifley had recommended some years before in a minority report written when he was a member of the Royal Commission on Monetary and Banking Reform (1935–6)), and in a controversial move used troops to work open-cut mines during the *Coalfields Strike of 1949. Chifley's description of the goal of the labour movement as the 'light on the hill' has become famous. In a speech to the New South Wales state conference of the *Australian Labor Party in June 1949 he declared: 'We have a great objective—the light on the hill—which we aim to reach by working for the betterment of mankind not only here but anywhere we may give a helping hand'. Chifley led the Opposition from 1949 until his death.

CHISHOLM (née Jones), CAROLINE (1808–77), philanthropist, was born in England. At the age of twenty-two she married Archibald Chisholm of the East India Company, and converted to Roman Catholicism at about this time. In 1832 she founded the Female School of Industry for the Daughters of European Soldiers, in Madras. From 1838 to 1846 she lived in Australia, where she founded a Female Immigrants' Home in Sydney in 1841. She accompanied immigrants to country areas to arrange accommodation and employment for them; organized a settlement for about twenty-three families during the *depression of the 1840s; and she and her husband travelled throughout New South Wales and collected about six hundred statements from immigrants about their lives in Australia. From 1846 to 1854 Caroline Chisholm lived in England. She gave evidence before two House of Lords committees, one on the administration of the criminal law, the other on colonization from Ireland; she disseminated much information about Australia; and in 1849 she founded the Family Colonization Loan Society, which initially chartered ships to bring settlers to Australia. From 1854 to 1866 she again lived in Australia, and was active in public life until ill health curtailed her activities. In 1854 she toured the Victorian goldfields, after which she arranged for the construction of shelter sheds on the roads to the diggings. She campaigned against the *squatters' hold on the land. In 1862, because of financial difficulties, she opened a girls school in Sydney, then returned to England in 1866 and lived there until her death. She wrote a number of pamphlets.

CHURCH AND SCHOOL CORPORATION
A body established in New South Wales in 1826, under instructions from the British Government, which was to be given one-seventh of the colony's lands, with which to maintain the Church of England's churches and schools. Other religious denominations opposed its existence. The Corporation received some lands and financial support, but its operations were suspended in 1829, and it was finally dissolved in 1833.

CLARK, ANDREW INGLIS (1848–1907), lawyer and politician, was born and educated in Hobart. He worked in his family's engineering firm before studying law, and was admitted to the Bar in

J. W. Beattie's photograph, Andrew Inglis Clark the elder, *from J. W. Beattie's* Members of the Parliaments of Tasmania *(c. 1900).* Clark *played a major role in the federation movement, and is also remembered for having introduced the Hare–Clark system. (Allport Library and Museum of Fine Arts, State Library of Tasmania)*

1877. Clark was a Member of the Tasmanian House of Assembly for some years (1878–82, 1887–98), during which time he was Attorney-General twice (1887–92, 1894–7), and initiated much legislation, including that of 1896 introducing the *Hare–Clark system of voting (as it was later called) in Tasmania. He attended the Federal Council several times, the Australasian Federation Conference of 1890, and the National Australasian Convention of 1891 (during which he helped *Griffith, *Barton and *Kingston to produce the draft constitution adopted by the convention, having already written his own draft constitution before the convention). His knowledge of, and admiration for, the American system were evident. Clark was appointed to the Tasmanian Supreme Court in 1898. He helped to found the University of Tasmania, and was its Vice-Chancellor from 1901 to 1903. His publications include *The Federal Financial Problem and Its Solution* (1900) and *Studies in Australian Constitutional Law* (1901).

CLARK, CHARLES MANNING HOPE (1915–91), historian, was born in Burwood, Sydney, the son of a Church of England minister and his wife, a descendant of Samuel *Marsden. Clark was educated at Belgrave State School, Mont Albert Central, and the University of Melbourne, from which he graduated with an MA. He studied at Oxford University in 1938 and taught at Blundell's School in England in 1939 before returning to Australia in 1940. He taught history at Geelong Grammar School from 1940 to 1944, lectured in political science at Melbourne University from 1944 to 1946, and then lectured in history at that university from 1946 to 1948, offering the first full Australian history course at any Australian university in 1946. Manning Clark became the first Professor of Australian History at the Canberra University College, later part of the Australian National University (ANU), in 1949. He remained at the ANU until his retirement in 1975. His *Select Documents in Australian History 1788–1850* (1950), *Select Documents in Australian History 1857–1900* (1955), and *Sources of Australian History* (1957) contributed much to the developing study of Australian history, but his chief work is his widely-read and controversial six-volume narrative history entitled *A History of Australia* (1962–87), which the historian Bill Gammage describes as 'a national treasure, and a contributor to the discourse of humanity'. Professor Clark, Australia's best-known historian, frequently spoke out on subjects ranging from cricket to republicanism until his death in Canberra. His other publications include *Meeting Soviet Man* (1960), the result of a visit to Russia in 1958–9; the popular *A Short History of Australia* (1963); a collection of autobiographical short stories entitled *Disquiet and Other Stories* (1968); a biography, *In Search of Henry Lawson* (1978); and a two-volume autobiography, *The Puzzles of Childhood* (1989) and *The Quest for Grace* (1990).

CLARKE, MARCUS ANDREW HISLOP (1846–81), journalist and novelist, was born in London. His father was a lawyer. Clarke migrated to Australia in 1863, and held various jobs. He worked for the Bank of Australasia; he worked on a station in Victoria; he contributed to various journals and newspapers, including the Melbourne *Argus* (on whose staff he worked for a time), the Melbourne *Herald*, the London *Daily Telegraph* and the Melbourne *Age*; he owned

An advertisement for the film For the Term of His Natural Life *(1908), an adaptation of a stage play based upon Marcus Clarke's novel of the same name. The film, a great hit, included some scenes which were filmed in the ruins of Port Arthur. (Archives Office of Tasmania)*

and edited the *Colonial Monthly* for some time; he edited the *Humbug* and the *Australian Journal*; and he was secretary to the trustees (1870–3), and assistant librarian (1873–81), of the Public Library of Victoria. Clarke is best known for his novel *His Natural Life* (1874), usually called by its later title, *For the Term of His Natural Life*. First published in the *Australian Journal* (1870–2), it tells the story of a convict's life in Van Diemen's Land. Clarke's numerous other publications include *Old Tales of a Young Country* (1871), *The Future Australian Race* (1877), *Sensational Tales* (1886), and *Stories of Australia in the Early Days* (1897). He also wrote essays and drama.

CLARKE, WILLIAM JOHN TURNER (1801?–74), pastoralist, was born in England, where he worked as a drover for some time, before migrating to *Van Diemen's Land in 1829. He worked as a butcher and government meat contractor, and began to acquire land in Van Diemen's Land, and later in the *Port Phillip District. Clarke built up large flocks of sheep and introduced the Leicester breed to Australia. He was a Member of the Victorian Legislative Council (1856–61, 1863–70), and a director of the Colonial Bank. 'Big' Clarke, as he came to be called, was reputedly the wealthiest man in the country, and left approximately £2 500 000 and much property in Tasmania, Victoria, South Australia and New Zealand. Other members of Clarke's family, notably his son Sir William John Clarke (1831–97), a pastoralist and philanthropist, have also been prominent in Victoria's history.

CLOSER SETTLEMENT Land policy pursued throughout Australia for many years, especially during the period from the 1880s to the 1920s. Its main aim was to establish small settlers on farms in specific areas, such as the Darling Downs in Queensland and Gippsland in Victoria, but it proved very expensive. The reasons for closer settlement included the failure of many *selectors; the need for more agricultural production because of Australia's increased population; improved agricultural, transport, and refrigeration methods; and the interest in land policy stimulated by the theory of Henry George, an American economist, that there should be a single tax on land. Governments opened up new areas of Crown land, introduced land taxes, compulsorily acquired land from large land-owners and made it available for smaller settlers, expanded railways, supported irrigation schemes such as those begun by the *Chaffey brothers, and provided credit and other assistance. Closer settlement was sometimes linked with migration schemes. Group settlement, involving groups of settlers working together, and *soldier settlement, were specialized forms of closer settlement.

CLUNIES ROSS, SIR IAN (1899–1959), scientist, was born at Bathurst, New South Wales, and educated at Newington College, Sydney, and the University of Sydney, graduating with a Bachelor of Veterinary Science degree with honours in 1921. After a year working in London as the Walter and Eliza Hall Veterinary Science Fellow, he returned to Sydney University as a Lecturer in Veterinary Science. Clunies Ross worked for the Council for Scientific and Industrial Research (CSIR) from 1926 to 1937, then chaired the International Wool Secretariat from 1937 to 1940. In 1938 he was a member of the Australian delegation to the *League of Nations Assembly. He was Professor of Veterinary Science at Sydney University from 1940 to 1946, and President of the Australian Institute of International Affairs from 1941 to 1945. Clunies Ross was a member of the CSIR's executive committee from 1946 to 1949, and the first chairman of its successor, the *Commonwealth Scientific and Industrial Research Organization (CSIRO), from 1949 until his death. Under his chairmanship, the CSIRO did much to assist the Australian sheep and wool industry. His publications include (with H. M. Gordon) *The Internal Parasites and Parasitic Diseases of Sheep: Their Treatments and Control* (1936) and many research papers.

CLYDE COMPANY   A pastoral company formed in 1836 in Glasgow, Scotland, by five local merchants and two men from the Clyde River in *Van Diemen's Land, Patrick Wood and Philip Russell, to raise stock in the *Port Phillip District (later Victoria). In 1836 the Company's manager (and a partner from 1841), George Russell, Philip's half-brother, settled on land on the Moorabool River, near present-day *Geelong. Three years later he transferred the Company's headquarters to Leigh Creek, further to the west. By late 1840 the Company was running 11 000 sheep and 300 cattle on some 8100 hectares, most of the land being leased. Despite the *depression of the 1840s, the Company prospered, acquired more land further to the west at Terinallum on Mount Emu Creek in 1846, and was running some 60 000 sheep by 1848. The Company was dissolved in 1858.

COALFIELDS LOCK-OUT   A lock-out (that is, 'a refusal on the part of an employer, or employers acting in concert, to furnish work to their operatives, except on conditions to be accepted by the latter collectively') by mine owners on the northern coalfields of New South Wales from March 1929 to June 1930. Mine owners locked out some 10 000 miners after the miners refused to accept reduced wages. The New South Wales Government attempted to reopen the Rothbury mine in December 1930; one miner, Norman Brown, was killed and nine others were wounded in a violent clash between miners and police. The miners returned to work, basically on the mine owners' terms.

COALFIELDS STRIKE A major industrial dispute in 1949. Some 23 000 black-coal miners throughout Australia went on strike on 27 June. The Communist-led Miners' Federation was demanding long-service leave, a 35-hour week, and increased wages. The effects of the strike were widespread; heavy restrictions on the use of coal and gas were imposed, industry almost ceased, and unemployment soared. The *Chifley Labor Government legislated quickly to freeze union funds, and eight union officials were jailed under this legislation. Australian Railways Union members, under heavy police protection, shifted coal from some coalfields. Chifley announced that troops would be used to work open-cut mines, and some 2500 troops began to do so on 1 August. Their demands had not been met when the miners returned to work on 15 August. There has been much debate about the causes and handling of the strike.

COBB & CO. Famous Australian coaching firm. Four Americans (Freeman Cobb, John B. Lamber, John Murray Peck, and James Swanton) founded a carrying firm in 1853 with wagons running from Melbourne to Liardet's Beach (present-day Port Melbourne). Cobb & Co. began taking passengers to the Victorian gold-fields in 1854. The syndicate sold the business in 1856, and several different owners followed. James Rutherford, an American who was to play an important role in running Cobb & Co. over the next fifty years, headed a syndicate that bought the business in 1861. Cobb & Co. expanded to New South Wales in 1862 and Queensland in 1865. Bathurst, New South Wales, became the company's headquarters in 1862. Cobb & Co. is best known for its coaching activities: in 1870 it used 6000 horses daily, its coaches covered 28 000 miles weekly, and its mail subsidies were £95 000 yearly. The company also had manufacturing and pastoral interests. The last Cobb & Co. coach ran in Queensland in 1924, but the name Cobb & Co. has become part of Australian folklore.

COLES, SIR GEORGE JAMES (1885–1977), business man, was born at Jung, Victoria, and educated at Beechworth College. In 1914 he opened a variety shop in Collingwood, the slogan of which was 'Nothing over a shilling', but sold it in 1916 and served in the first AIF until 1918. (His brother James, who partnered him in the business, was killed during the First World War.) George and another brother, Arthur (later Sir Arthur) William Coles, opened another variety shop in Collingwood in 1919. Its slogan, 'Nothing over half-a-crown' (two shillings and sixpence), remained in force until 1938. A company was established in 1921, with George as its managing director until 1931, followed by Arthur from 1931 to 1940. Sir Kenneth Frank Coles, Sir Edgar Barton Coles and Sir Norman Cameron Coles, all brothers of George and Arthur, also became involved in the business. G. J. Coles & Coy Ltd developed into a very large retail business, and merged with the Myer organization (which had been founded by *Myer) in 1985.

COLLINS, DAVID (1756–1810), colonial administrator, was born in England. Like his father, he pursued a career in the marines. Collins came to New South Wales as the colony's Deputy-Judge-Advocate with the *First Fleet. He remained in the colony in that position until 1796, when he returned to England. Collins was commissioned as Lieutenant-Governor of a proposed new British convict settlement in the Bass Strait area in 1803, unsuccessfully attempted to establish such a settlement near present-day *Sorrento later that year, and finally transferred the settlement to *Van Diemen's Land in 1804. He was the first lieutenant-governor of Van Diemen's Land from then until his death. His publications include *An Account of the English Colony in New South Wales* (two volumes, 1798, 1802).

'COLLINS, TOM' *Furphy, Joseph

COLOMBO PLAN   A plan designed to foster economic development within the South and South-East Asian region. It originated from discussions at a conference of foreign ministers of the *Commonwealth of Nations held in Colombo in 1950. Donor countries were to provide financial, technical, and educational assistance. Members eventually came to include Australia, Canada, Japan, New Zealand, the UK, the USA, and twenty-one countries within the region.

COLONIAL SUGAR REFINING COMPANY LIMITED (CSR)   One of Australia's largest companies, formed in Sydney in 1855, originally to refine raw sugar. It began to manufacture raw sugar in New South Wales in 1870, and eventually operated a large number of mills and plantations, many of which were in Queensland. It also operated in Fiji and New Zealand. From the late 1930s onwards it diversified into other industries, and became known as CSR Limited in 1973.

COMMONWEALTH BANKING CORPORATION   A statutory banking organi-

The Colonial Sugar Refining Company Limited's Victoria Mill seen here in its first year of operation, 1883. The sugar mill, Australia's largest, crushed 2 370 748 tonnes of cane in 1989. (Archives of Business and Labour, Australian National University)

zation. The *Fisher Labor Government introduced legislation to establish the Commonwealth Bank, sometimes referred to as the 'people's bank', in 1911. It was to be a savings and trading bank, capitalized by the Commonwealth Government, and competing with other banks. Its savings-bank functions began in 1912 (the Commonwealth Savings Bank of Australia being created in 1927), and its trading bank functions began in 1913 (the Commonwealth Trading Bank being created in 1953). The Commonwealth Bank began to operate as a central bank from about 1930. *Chifley increased its powers in 1945, but his attempt to introduce *bank nationalization in 1947 failed. In 1959 legislation was passed to establish the *Reserve Bank of Australia, which was to take over central banking functions, and the Commonwealth Banking Corporation, which was to comprise the Commonwealth Savings Bank, the Commonwealth Trading Bank, and the new Commonwealth Development Bank of Australia. The Commonwealth Banking Corporation came into being in 1960.

COMMONWEALTH CONCILIATION AND ARBITRATION COURT Federal body established under the *Commonwealth Conciliation and Arbitration Act* 1904 to settle interstate industrial disputes. The so-called *Harvester Judgement of 1907 was one of the Court's best-known decisions. In 1956, following a *High Court decision, the Court was replaced by two new bodies, the Commonwealth Conciliation and Arbitration Commission (the Australian Industrial Relations Commission from 1989) and the Commonwealth Industrial Court. The Federal Court of Australia took over the latter body's work in 1977.

COMMONWEALTH GRANTS COMMISSION Statutory body established in 1933 to give advice on applications by the states to the Commonwealth for special financial assistance. The original claimant states were South Australia (until 1959, and again between 1970 and 1975), Tasmania (until 1974) and Western Australia (until 1968), the original standard (non-claimant) states being Victoria, New South Wales and Queensland (which many years later became a claimant state). A proposal in 1936 to abolish the Commission was abandoned in 1938. Following the introduction of *uniform taxation in 1942, the Commission gave advice on additional tax reimbursement grants (as well as special grants) until 1946, after which it continued to advise on special grants. Tax reimbursement grants were replaced by financial assistance grants in 1959, which in turn were replaced by a system of personal income tax sharing in 1976, the Commission continuing throughout to give advice on special grants. It gained some powers (later changed) to advise local government about financial assistance in 1973, and powers to advise on matters relating to the Northern Territory in 1978 and the Australian Capital Territory in 1980. It began reviews of tax-sharing relativities in 1978. The Commission was located in Melbourne until 1972, when it was moved to Canberra.

COMMONWEALTH OF AUSTRALIA Official title of the Australian nation. It comprises the six states that were formerly colonies—*New South Wales, *Queensland, *South Australia, *Tasmania, *Victoria, and *Western Australia—and the *Australian Capital Territory and the *Northern Territory. The *federation movement occurred mainly during the 1880s and 1890s. The

*Federal Convention (1897–8) produced a draft federal constitution, which was put to the people at two referendums. The first failed; the second, which was held after amendments were made, passed in all of the colonies. This constitution formed the basis of the *British Commonwealth of Australia Constitution Act* 1900. The Commonwealth of Australia was inaugurated on 1 January 1901. Powers were distributed between the Commonwealth and state parliaments; the former consists of the House of Representatives and the Senate. In 1911 the Northern Territory was transferred from South Australia to the Commonwealth and land was transferred from New South Wales to the Commonwealth for the creation of the Australian Capital Territory. The Commonwealth of Australia is not to be confused with the *Commonwealth of Nations, to which Australia belongs.

COMMONWEALTH OF NATIONS An association of nations, including Australia, which were formerly part of the British Empire. The term 'British Commonwealth' began to replace 'British Empire' during the *First World War. The title was changed to the Commonwealth of Nations in the late 1940s. The British monarch is the head of the Commonwealth of Nations.

COMMONWEALTH SCIENTIFIC AND INDUSTRIAL RESEARCH ORGANIZATION (CSIRO) A statutory body, which replaced the Council for Scientific and Industrial Research in 1949. The latter body, which had replaced the Institute of Science and Industry in 1926, had devoted most of its research to primary industries, but it had also assisted secondary industries from 1936 onwards. The CSIRO's many functions include research, training of research workers, liaison with research institutions in Australia and other countries, and publication of scientific and technical material. Its achievements have included the introduction of myxomatosis to reduce rabbit numbers, developments in the control of diseases affecting livestock, advances in the processing of textiles, and significant research on soils. Its first chairman was Ian *Clunies Ross.

COMMUNIST PARTY OF AUSTRALIA An Australian political party. J. S. Garden and W. P. Earsman were largely responsible for founding it at a meeting of twenty-six socialists in Sydney in 1920. It affiliated with the Comintern in 1922. During the Second World War it was banned, under wartime powers, from 1940 to 1942. In terms of membership and influence in the trade union movement, it reached its peak during the 1940s. F. W. Paterson, its only parliamentarian, was a Member of the Queensland Legislative Assembly from 1944 to 1950. *Menzies's attempts to dissolve the Communist Party in the early 1950s failed; the *Communist Party Dissolution Act* 1950 was found to be invalid by the *High Court in 1951, and a referendum that would have given the Commonwealth Government the power to ban communism was narrowly defeated on 22 September 1951. The party suffered two major splits; the first led to the formation of the Peking-oriented Communist Party of Australia (Marxist–Leninist) in 1963, and the second to the formation of the Moscow-oriented Socialist Party of Australia in 1971. The Communist Party of Australia was disbanded in 1989.

COMPULSORY VOTING The obligation to vote at political elections. The principle of compulsory voting was introduced in Queensland in 1915, Victoria in 1926, New South Wales and Tasmania in 1928, Western Australia in 1936, and South Australia in 1942. The same principle was introduced for Commonwealth elections in 1924. It is now compulsory to vote in elections for all Australian houses of parliament, except the South Australian Legislative Council.

CONDER, CHARLES EDWARD (1868–1909), painter, was born and educated in England. His father was a railway engineer. Conder arrived in New South Wales in 1884 and worked in the office of the Lands Department in Sydney for about eight months, in trigonometrical survey camps in New South Wales for about two years, and then as an illustrator for the *Illustrated Sydney News*. He attended night classes at the Royal Art Society of New South Wales, then, from 1888 to 1890, at the Melbourne National Gallery School. Conder became one of the leading painters in the group that became known as the *Heidelberg School, and contributed more than forty works to the controversial *Exhibition of 9 x 5 Impressions*. In 1890 Conder moved to Paris, where he studied at the Atelier Cormon for some time, and in 1894 he moved to London. He became famous for his watercolours on silk fans. He died in England. His Australian paintings include *The Departure of the S.S. Orient Circular Quay* (1888), *A Holiday at Mentone* (1888), and *Cove on the Hawkesbury* (c. 1888).

CONISTON STATION A pastoral station in the Northern Territory, some 255 kilometres north-west of Alice Springs, which was the scene of the last punitive expeditions against Aborigines. In 1928 members of the Walbiri people had moved closer to the station because of severe drought conditions. On 7 August, apparently acting in response to the abduction of an Aboriginal woman, some of them killed a White dingo-hunter named Fred Brooks. The Government Resident of Central Australia, C. A. Cawood, dispatched a police expedition, led by Mounted Constable William Murray, to find those responsible. The expedition killed a number of Aborigines and brought back two others, who were tried in Darwin for murder, but were acquitted. Another police expedition, also led by Murray, was sent out following another Aboriginal attack on 28 August, apparently for similar reasons, on White pastoralist 'Nugget' Morton, who survived. After the two expeditions, which killed some 70 to 100 Aborigines, not 31 as the official figures claimed, the Walbiri people fled from their land. Public outrage throughout Australia followed the killings. The *Bruce–*Page Government held a Commission of Inquiry, whose members included Cawood, which found in January 1929 that the police action was justified.

CONSCRIPTION Compulsory enlistment for military service. The *Defence Act 1903* introduced the principle of conscription of men for service within Australia during wartime, but not for overseas service during either wartime or peacetime. The *Defence Act 1909* introduced the principle of conscription of men for service within Australia during peacetime. Compulsory military training for boys and men, aged between 12 and 25, was introduced in 1911. This system, which was conducted on a part-time basis, and was restricted to home

Conscription

In the 1916 conscription referendum 'The Blood Vote', a leaflet with words by W. R. 'Winspear' and cartoon by Claude Marquet, urged a 'No' vote. The devil in the background, complete with cloven hoof, is meant to be Billy Hughes. (Australian War Memorial)

service, continued during the First World War. The Commonwealth Government, led by *Hughes, made two unsuccessful attempts to introduce conscription of men for overseas service during the war. Two referendums (or plebiscites) on the subject were held, the first in 1916, the second in 1917, amidst intensely bitter debates. Both failed. After the First World War, compulsory military training continued, although on a diminishing scale during the 1920s. The *Scullin Labor Government abolished it in 1929. The *Menzies *United Australia Party Government reintroduced conscription of men for service within Australia in 1939, and in 1943 the *Curtin Labor Government introduced a limited form of conscription of men for overseas service.

Conscripts could be sent to a prescribed area, the 'South-Western Pacific Zone', outside Australia. (A form of industrial conscription for men and women had also been introduced in Australia during the Second World War.) The Menzies Liberal–Country Party Government reintroduced conscription of men in 1951. This became known as the National Service scheme. Eighteen-year-olds were required to complete a period of training (part of which was full-time), and then to remain in the Reserve for some years. Until 1957 they could choose to serve in the navy, the army, or the air force. Those who chose the services other than the army had to be prepared to serve overseas. From early 1957 conscripts served in the army and were selected by ballot, but this scheme was abolished in 1959. The Menzies Liberal–Country Party Government introduced another National Service scheme in 1965. Twenty-year-old men were selected by ballot to serve in the army for two years on a full-time basis, and they could be sent overseas. This scheme was in operation when the *Holt Liberal–Country Party Government decided to send conscripts to the *Vietnam War in 1966. Almost 64 000 men were conscripted under this scheme (some others chose alternatives, such as part-time service in the citizen forces) from 1966 to 1972; of these some 15 500 served in Vietnam. There were intensely bitter conscription debates again, and the period of service was reduced to eighteen months in late 1972. The newly elected *Whitlam Labor Government abolished conscription in December 1972.

CONSTITUTION OF THE COMMONWEALTH OF AUSTRALIA Australia's written constitution. It is the ninth clause of the *British Commonwealth of Australia Constitution Act* 1900. This Act created a federation, of which the colonies became states. The constitution consists of 128 sections, divided into eight chapters: The Parliament, The Executive Government, The Judicature, Finance and Trade, The States, New States, Miscellaneous, and Alteration of the Constitution. A referendum must be carried to alter the Constitution.

CONVICTS *Transportation

COOK, JAMES (1728–79), navigator and cartographer, was born in England, educated at a village school, apprenticed to a shopkeeper for about eighteen months, and then worked for a coal-shipper for about three years. In 1755 he joined the British navy. Cook is best known for three expeditions. As Lieutenant, he commanded HM Bark *Endeavour* on an expedition to observe the transit of Venus and to explore the South Pacific (1768–71). *Banks accompanied him on this expedition. After observations were made at Tahiti (1769), Cook circumnavigated New Zealand, charted its coast, and claimed possession of some areas for Britain. In 1770 he sailed north along the eastern coast of *New Holland, made several landings, including one at *Botany Bay, and claimed possession of the eastern part of the continent (an area he later named *New South Wales) for Britain on 22 August 1770. He confirmed the existence of Torres Strait. From 1772 to 1775 Cook was in charge of an expedition to discover whether there was another southern continent. He sailed in the *Resolution*, with Tobias Furneaux in the *Adventure*. Cook circumnavigated the world in high southern latitudes, but did not find another continent. He discovered New Caledonia and *Norfolk Island. Furneaux sailed along the southern and eastern coasts of *Van Diemen's Land during this voyage. As Post-Captain, and a Fellow of the Royal Society, Cook commanded an expedition to search for a passage between the North Pacific and the North Atlantic; he set sail in the *Resolution*, with Charles Clerke in the *Discovery*, in 1777. During the voyage Cook visited Van Diemen's Land. He explored the Pacific coasts of North America and Siberia, but failed to find a passage. Cook was killed by islanders in the Sandwich Islands (present-day Hawaii), and the *Resolution* and the *Discovery* returned to England in 1780.

COOK, SIR JOSEPH (1860–1947), politician, was born in Staffordshire, England, left school at about nine, and worked in the mines. In 1885 he migrated to New South Wales where he worked as a miner, and later as an auditor, and continued his trade-union involvement, begun in England. He was a Member of the New South Wales Legislative Assembly from 1891 to 1901, elected first to represent Labor, but after refusing to sign the solidarity pledge in 1894, he stood as an Independent Labor candidate. Cook was a Member of the House of Representatives from 1901 to 1921, becoming Leader of the Free Traders in 1908. He became Deputy Leader and Minister for Defence in the *Deakin Fusion Government (1909–10). Cook was largely responsible for the defence legislation changes providing for a compulsory military training scheme for males within Australia and the establishment of a military college; he concluded naval arrangements with Britain, and also played a large part in arranging Kitchener's visit to Australia. At the 1913 federal elections, he led the Liberals (formerly the Fusionists); they narrowly won control of the House of Representatives, but not of the Senate, and Cook was Prime Minister from June 1913 to September 1914. Labor won the 1914 elections, following a double dissolution. When the Nationalist Party emerged in 1917, Cook became Deputy Leader. He was Australian High Commissioner in London from 1921 to 1927.

COPLAND, SIR DOUGLAS BERRY (1894–1971) economist and diplomat, was born at Timaru, New Zealand, and graduated from the University of New Zealand. He was Professor of Economics at the University of Tasmania from 1920 to 1924, then Sidney Myer Professor of Commerce and Dean of the Commerce Faculty at the University of Melbourne from 1924 to 1944. During the *Great Depression of the 1930s he chaired the committee that was largely responsible for formulating the so-called Premiers' Plan. Copland was Commonwealth Prices Commissioner (1939–45), and also an economic consultant to the Prime Minister (1941–5) during the Second World War. He then served as Australian Minister in China (1946–8), the first Vice-Chancellor of the Australian National University (1948–53), Australian High Commissioner to Canada (1953–6), and the first principal of the Australian Administrative Staff College at Mount Eliza (1956–9). He also chaired the newly formed Committee for the Economic Development of Australia from 1960 to 1966. His numerous publications include *Australia in the World Crisis, 1929–1933* (1934), *W. E. Hearn: First Australian Economist* (1935) and *The Changing Structure of the Western Economy* (1961).

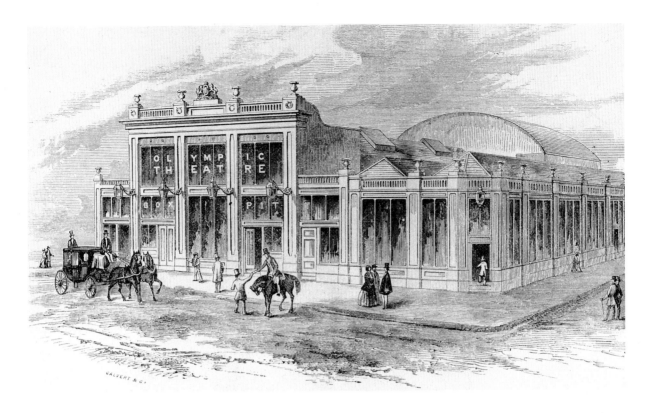

'COPPER CITY' *Mount Isa

COPPIN, GEORGE SELTH (1819–1906), comic actor and entrepreneur, was born in Sussex, England, the son of strolling players, and became involved in the theatre at an early age. He arrived in Sydney in 1843. Coppin, who has frequently been described as 'the father of Australian theatre', acted in most of the Australian colonies and New Zealand; he founded several theatre companies; he brought more than two hundred overseas entertainers, including *Williamson and Maggie Moore, to Australia; and he built a number of theatres, the best-known being the prefabricated Olympic Theatre (the so-called 'Iron Pot') in Melbourne. His other business interests included hotel-keeping, mining and racing. Coppin also had a political career in Victoria: he was a Member of the Legislative Council (1858–63, 1889–95), and of the Legislative Assembly (1874–7, 1883–8), and was responsible for introducing the *Real Property Act* 1862, based on the *Torrens system. His philanthropic work included helping to found the Victorian Humane Society, and he was an active Freemason.

COSME *New Australia; *Lane, William

COUNCIL FOR SCIENTIFIC AND INDUSTRIAL RESEARCH *Commonwealth Scientific and Industrial Research Organization

COUNTRY PARTY *National Party of Australia

COUNTRY WOMEN'S ASSOCIATION (CWA) A non-political, non-sectarian, voluntary organization. Open to all women, the CWA was founded in New South Wales

*Engraving by Samuel Calvert of **Coppin**'s Olympic Theatre, probably for a letterhead. (La Trobe Collection, State Library of Victoria)*

and Queensland in 1922, Western Australia in 1924, Victoria in 1928, South Australia in 1929, Tasmania in 1936, and the Northern Territory in 1960. The national organization was founded in Melbourne in 1945. The CWA provides educational, cultural, health and other services, and has links with the Associated Country Women of the World.

COWAN (née Brown), **EDITH DIRCKSEY** (1861–1932), social worker, politician, and feminist, was born near Geraldton, Western Australia, and educated in Perth. Her father was a pastoralist and her mother was a teacher. She became active in numerous organizations, some of which she had helped to found, including the Karrakatta Women's Club, of which she was variously secretary, vice-president, and president; the Women's Service Guild, of which she was the original vice-president from 1909 to 1917; and the Western Australian branch of the *National Council of Women, of which she was president from 1913 to 1921, then vice-president until her death. She helped to found the Children's Protection Society in 1906, was appointed to the bench of the Children's Court in 1915, and became a Justice of the Peace in 1920, the year she was awarded an OBE for her war work with the Red Cross and other organizations. Edith Cowan made several overseas trips, including one to the United States in 1925, where she represented Australia at the seventh International Conference of Women. In 1921 she became the first woman member of any Australian parliament, when she was elected as a Nationalist Member of the Western Australian Legislative Assembly. She argued for reforms relating to the treatment of women, infants, and migrants, and introduced the *Women's Legal Status Act* 1923, which lifted restrictions on the entry of women to civil office and the legal profession. She was defeated in the 1924 elections, and was unsuccessful when she stood again in 1927. An active member of the Church of England, Edith Cowan died in Perth.

COWPER, SIR CHARLES (1807–75), politician, was born in England, and brought to Australia in 1809. His father was a clergyman. Cowper was privately educated. He worked in the Commissariat Department for some time; was the secretary of the *Church and School Corporation from 1826; became a pastoralist; and managed the Sydney Tramway and Railway Company for some years from 1849. He was a Member of the Legislative Council of New South Wales for several periods (1843–50, 1851–6, 1860), and played a leading role in the movement against *transportation in the late 1840s and early 1850s. Cowper was also a Member of the Legislative Assembly of New South Wales for most of the time from 1856 to 1870. He was Premier five times (1856, 1857–9, 1861–3, 1865–6, 1870) and Colonial Secretary six times (1856, 1857–9, 1860–1, 1861–3, 1865–6, 1870). He played a large part in New South Wales' adoption of manhood *suffrage in 1858, introduction of land reforms in 1861, abolition of state grants to religion in 1862, and adoption of the *Torrens system in 1862. After resigning from politics in 1870, he became the Agent-General for New South Wales in London, a position he held until his death.

COWRA OUTBREAK An attempted escape in 1944 by about 1100 Japanese prisoners-of-war from a camp at Cowra, New South Wales, during the Second World War. They used makeshift weapons, such as baseball bats, and several hundred prisoners escaped, but were killed or recaptured. Four guards and more than 230 prisoners were killed; three guards and more than 100 prisoners were wounded.

CRESSWELL, SIR WILLIAM ROOKE (1852–1933), vice-admiral, was born at Gibraltar, where his father was Deputy Postmaster-General. He was educated at Aitken's Private School, Gibraltar, and Eastman's Naval Academy, Southsea, England, and entered the Royal Navy as a cadet in 1865. He served in various parts of the world before resigning in 1878. Cresswell arrived in Australia in 1879, and engaged in droving and other activities in the Northern Territory until 1885, when he entered the South Australian defence forces, becoming Naval Commandant in 1893, and being promoted Captain in 1895. He became Commandant of the Queensland naval forces in May 1900. During the *Boxer Rising, he commanded the *Protector* (his former ship from South Australia) from August 1900 to January 1901. A strong advocate of the creation of an Australian navy, Cresswell became Director of the Commonwealth naval forces in 1904 (during which year he had also become Naval Commandant of Victoria), and a member of the Council of Defence, and of the Australian Naval Board, in 1905. He became Rear-Admiral, and First Naval Member of the Commonwealth Naval Board in 1911. Cresswell played a

Cricket was well established in the Australian colonies by the time This Sketch of the Victorian Eleven and the Intercolonial Cricket Ground, Melbourne, Feby. 2-3 & -4-1860 *by E. Gilks was published in 1860. (La Trobe Collection, State Library of Victoria)*

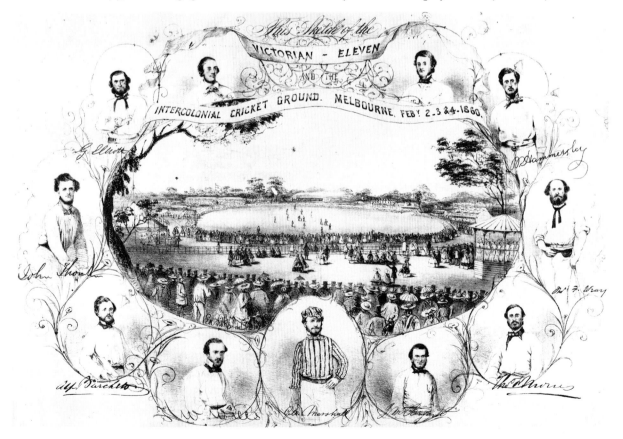

large role in the development of the *Royal Australian Navy before and during the First World War. He retired in 1919, and was promoted to Vice-Admiral in 1922.

CRICKET was introduced to New South Wales at least as early as 1803, and has become one of Australia's most popular sports, played mainly by men. The Melbourne Cricket Club, which was to contribute significantly to the development of Australian cricket, was founded in 1838. Test matches have been played against England since 1876–7, the first being held in Australia. (An Aboriginal team had earlier toured England, in 1868.) After an Australian team won a match in London in 1882, the London *Sporting Times* published a notice that English cricket had died, that the body was to be cremated, and that the ashes were to be taken to Australia. When an English team won a series held in Australia in 1882–3, the English captain was given an urn of ashes, which was taken to Lord's Cricket Ground in London, where it remains, regardless of which team 'wins the Ashes', that is, wins the greater number of matches in a Test series. The Australian Board of Control for International Cricket was founded in 1905. The *bodyline series of 1932–3 was particularly controversial. Australia has also played Test matches against South Africa (from 1902–3 to 1969–70), the West Indies (from 1930–1), New Zealand (from 1945–6), India (1947–8), and Pakistan (from 1956–7). The establishment of World Series Cricket by Kerry Packer in 1977 was controversial. Australia's domestic first-class competition, played for the Sheffield Shield Trophy, began in 1892–3. Notable Australian cricketers have included members of the famous Gregory family, Frederick Robert Spofforth (1853–1926), Montague Alfred Noble (1873–1940), Victor Trumper (1877–1915), Warwick Windridge Armstrong (1879–1947), Charles George Macartney (1886–1958), *Bradman (1908–), Keith Miller (1919–) and Richard Benaud (1930–).

'CRIMSON THREAD OF KINSHIP' A now-famous phrase used by *Parkes, the Premier of New South Wales, in a speech advocating the federation of the Australian colonies. He gave the speech at a dinner held during the Australasian Federation Conference in Melbourne in 1890. The full sentence read: 'The crimson thread of kinship runs through us all'.

CSIRO *Commonwealth Scientific and Industrial Research Organization

CSR *Colonial Sugar Refining Company Limited

'CULTURAL CRINGE' A term used to describe Australians' intimidation by, and subservience to, Anglo-Saxon culture. It was first used by the Australian critic A. A. Phillips as the title of an article published in *Meanjin* in 1950 and later reprinted in Phillips's *The Australian Tradition* (1958).

CURRENCY LADS AND LASSES The name given in the early nineteenth century to people of European descent born in Australia, especially those of convict parentage. British-born settlers were known as 'sterling'. The names came from the distinction made between 'currency', coins or promissory notes valid for local use only, and 'sterling',

British coins or treasury bills. 'Currency' was worth less than 'sterling'.

CURTIN, JOHN JOSEPH (1885–1945), politician, was born at Creswick, Victoria, where his father was a policeman. Curtin spent most of his early years in the Melbourne suburb of Brunswick. After leaving school at thirteen he worked in various jobs, mostly in the printing trade. He joined the Socialist Party in 1906, and later the Labor Party. He was secretary of the Timber Workers' Union from 1911 to 1915, then moved to Perth, where he edited the *West Australian Worker* from 1917 to 1928. He entered federal politics, and was a Labor Member of the House of Representatives (1928–31, 1934–45). During the *Great Depression, Curtin opposed both the Premiers' Plan and J. T. *Lang's plan. After losing his seat in 1931 he worked as a freelance journalist and as an advocate for Western Australia before the *Commonwealth Grants Commission. He led the Labor Opposition from 1935 to 1941, when he became Prime Minister, and held office until his death in July 1945, shortly before the end of the Second World War. Curtin insisted, in the face of opposition from Churchill and Roosevelt, on the return of Australian troops from the Middle East to defend Australia, rather than their diversion to defend Burma. He appealed to the USA, not Britain, for support, issuing a public statement (which appeared under his name, but was apparently written by his publicity officer) that included the following words: 'Without any inhibitions of any kind, I make it quite clear that Australia looks to America, free of any pangs as to our traditional links or kinship with the United Kingdom.' He strongly supported the American General Douglas MacArthur, Supreme Commander of the Allied Forces in the southwest Pacific, and in 1943 extended the area in which Australian conscripts could be used to include a number of places overseas, although he had been an anti-conscriptionist during the First World War.

CWA *Country Women's Association

CYCLONE TRACY, Australia's most destructive cyclone, struck *Darwin in the early hours of 25 December 1974. Wind speeds of 217 kilometres an hour were recorded in the city over a five-hour period, fifty people were killed, another sixteen were reported as missing, and most of the city's buildings were destroyed or damaged. The director of Australia's newly formed Natural Disasters Organization, Major-General Alan Stretton, supervised the evacuation of some 30 000 people from Darwin. The Darwin Reconstruction Commission organized the rebuilding of the city, which was largely completed by the end of 1977.

*John **Curtin**, Australia's Prime Minister from 1941 to 1945. (National Library of Australia)*

D

DAINTREE, RICHARD (1831–78), geologist and photographer, was born and educated in England, and arrived in Victoria in 1852, during the *gold-rushes. He was an assistant geologist in the Victorian Geological Survey from 1854 to 1856, went back to England and studied at the Royal School of Mines Laboratory for some time, and returned to Victoria in 1857. After working in the Geological Survey from 1859 to 1864, he settled on land in northern Queensland. Daintree was responsible for the opening of several Queensland gold-fields in the 1860s, before and during his term as government geologist in northern Queensland (1868–70). His photographs provide a valuable record of Queensland's early White settlement. He was a commissioner in charge of Queensland's display at the Exhibition of Art and Industry in London in 1871, and was then Queensland Agent-General in London from 1872 to 1876. He died in England. His publications include *Queensland, Australia: Its Territory, Climate, and Products* (c. 1872).

DAMPIER, WILLIAM (1652?–1715), mariner and writer, was born in England. He engaged in various trading, naval and privateering ventures. He and other buccaneers from the *Cygnet* spent about

*Richard **Daintree**'s Darley Quarries, Bacchus Marsh, Victoria (c. 1858). (National Gallery of Victoria)*

three months on the western coast of *New Holland (now Australia) in 1688. After returning to England in 1691 he wrote of his experiences in the South Seas. He was appointed to command an expedition to explore the eastern coasts of New Holland and New Guinea, and sailed from England in HMS *Roebuck* in 1699. He explored part of the western coast of New Holland and discovered and named New Britain, but the *Roebuck*'s poor state precluded exploration of the eastern coasts; it sank off Ascension Island on the return voyage. In 1702 Dampier was court-martialled over his conduct during the expedition and was found to be unfit to command naval vessels. In 1703 he led a disastrous privateering expedition to the South Seas in the *St George*, returning to England in 1707. He then sailed around the world between 1708 and 1711 as pilot on Captain Woodes Rogers's privateer, the *Duke*. Dampier's publications, which led to much interest in New Holland (even though they contained unfavourable comments), include *A New Voyage Round the World* (1697).

DARLING, SIR RALPH (1775–1858), *governor, was born to English parents in Ireland. He followed his father and pursued a military career. Darling succeeded *Brisbane as Governor of New South Wales in 1825, and was also nominally Governor of Van Diemen's Land, whose separate government he proclaimed in 1825, and which was administered by a lieutenant-governor. Darling's powers in New South Wales were limited by the Executive Council, which was established in 1825, and the Legislative Council, which was increased in size and authority in 1828. He was responsible for tightening up the convict system (as he had been instructed), and for improvements in the general administration of the colony. Like his predecessors, he faced problems because of the conflict between *emancipists and *exclusives in the colony. Darling's governorship ended in 1831. A British parliamentary inquiry into his actions in the *Sudds and Thompson affair exonerated him in 1835. He died in Brighton, England.

DARWIN, the administrative capital of the *Northern Territory, lies on the shores of Port Darwin. Officers of HMS *Beagle* discovered the harbour and named it after Charles Darwin in 1839. The settlement of Palmerston, on the site of present-day Darwin, began in 1862. It soon became known as Darwin, and its name was officially changed in 1911. Darwin was the northern base for the *Overland Telegraph Line. Troops were stationed there during the Second World War, and it became the target for repeated Japanese air attacks. The first and most severe of these occurred on 19 February 1942. The town was rebuilt after the war, and again after *Cyclone Tracy damaged or destroyed most of the city on 25 December 1974. The population in 1988 was 72 900.

DARWIN, BOMBING OF *Darwin; *Second World War

DAVIS, ARTHUR HOEY ('Steele Rudd') (1868–1935), writer, was born at Drayton, Queensland. His father was a blacksmith and *selector. Davis attended the Emu Creek school until he was about twelve years old, and worked as a public servant in Brisbane from 1885 to 1904. He was a regular contributor to the *Bulletin* from 1895 onwards. His several attempts to run his own journal resulted in *Steele Rudd's Magazine* (1903–7),

Steele Rudd's Annual (1917–23), *Steele Rudd's* (1924–5) and *Steel Rudd's and the Shop Assistant's Magazine* (1926–7). He also farmed from 1909 to 1917. Davis died in Brisbane. He is best known for his humorous descriptions of selection life, based partly on his own family's experiences, in books such as *On Our Selection* (1899), which has been adapted for stage, film (silent and sound) and radio (under the title of 'Dad and Dave'). His many publications include *Our New Selection* (1903), *Dad in Politics and Other Stories* (1908), and *The Rudd Family* (1926).

Davis Cup *Tennis

Dawson, Andrew (Anderson) (1863–1910), politician, was born in Rockhampton, Queensland. He left school at twelve and held various jobs, including that of miner, mostly in Charters Towers. He also became involved in the trade union movement and journalism. Dawson was a Labor Member of the Queensland Legislative Assembly from 1893 to 1901, and Premier and Chief Secretary in a minority Labor Government, which formally lasted from 1 December to 7 December 1899. It was the first Labor Government in any of the Australian colonies, and reputedly the first in the world. Dawson, who had supported federation, was a Labor Senator from 1901 to 1906, and Minister for Defence in the *Watson Labor Government of 1904. He was defeated, while standing as an Independent, in the federal elections of 1906. He died in Brisbane.

Deakin, Alfred (1856–1919), politician, was born in Melbourne, and educated at the Melbourne Church of England Grammar School and the University of Melbourne. He practised as a lawyer. After meeting David *Syme in 1878 he wrote for the Melbourne *Age*, and in 1880 edited the *Leader* (the *Age*'s weekly). From 1901 to 1914, in a secret arrangement, he was the special correspondent for the English *Morning Post*. He also wrote a number of books. Deakin was elected as a Liberal Member of the Legislative Assembly in Victoria in 1879, but lost his seat in a by-election held because of doubts about the earlier poll. After another unsuccessful attempt, he won a seat in 1880, held office in coalition governments from 1883 to 1890, and was Chief Secretary from 1886 to 1890. Deakin chaired a royal commission inquiring into *irrigation in 1884, visited California and met the *Chaffey brothers, and introduced the first legislation in Australia to promote an irrigation scheme. He remained on the back-benches for ten years from 1890 and also practised law again. He was active in various public organizations, such as the *Australian Natives' Association and the Imperial Federation League, and played a significant role in the *federation movement: he served on the Federal Council, attended the *Federal Convention, chaired the Federation League of Victoria, and took part in the delegation that went to London in 1900 to explain the Commonwealth of Australia Constitution Bill. From 1901 to 1913 he was a Member of the House of Representatives, Attorney-General in the *Barton Protectionist Government from 1901 to 1903, and succeeded Barton as Prime Minister and Minister for External Affairs in 1903, remaining in these positions until 1904. He was again Prime Minister, and Minister for External Affairs at the head of the Protectionists, from 1905 to 1908; and served as Prime Minister a third time, leading the so-called *Fusion Party in 1909–10. After its defeat by Labor, he led the Opposition (by then known as the Liberals) from 1910 to 1913, when he retired. His health, and his memory especially, had deteriorated before his death in 1919. Deakin's publications include *A New Pilgrim's Progress* (1877), *Irrigated India* (1893), and *Temple and Tomb in India* (1893).

DECIMAL CURRENCY Currency 'in which the units are decimal [of tenths or ten] multiples or fractions of each other'. Decimal currency was introduced in Australia on 14 February 1966. Dollars ($) (not Royals as *Menzies wanted them to be called) and cents replaced pounds (£) and pence. Coins of 1, 2, 5, 10, 20, and 50 cents were introduced, and notes of $1, $2, $10 and $20, the latter respectively replacing the 10 shilling, £1, £5, and £10 notes. Stuart Devlin designed the reverses of the new coins, and Gordon Andrews the new notes. The change-over, which was faster than expected, was virtually completed by the end of 1967. A $5 note was introduced in 1967, a $50 in 1973, and a $100 note in 1984. The $1 and $2 notes were replaced by coins in 1984 and 1988 respectively. The 1 cent and 2 cent coins were withdrawn in 1992. Historical figures represented on the notes have included John *Macarthur, William *Farrer, Sir Joseph *Banks, Caroline *Chisholm, Francis *Greenway, Henry *Lawson, Sir Charles *Kingsford Smith, Lawrence *Hargrave, Lord *Florey, Sir Ian *Clunies Ross, Sir Douglas *Mawson, and John *Tebbutt.

DEEMING, FREDERICK (BAILEY) (1853–92), murderer, was born in Leicestershire, England, the son of a grazier. Deeming, who worked as a plumber and gas-fitter and in a number of other jobs, lived variously in England, Australia, South Africa, and Uruguay. He served several gaol sentences for crimes such as theft and fraud. In March 1892 the body of Emily Lydia Mather, whom Deeming had married in 1891, was found cemented under a hearth-stone in a house in the *Melbourne suburb of Windsor. Some days later, Deeming was arrested in the Western Australian town of Southern Cross. Further investigations led to the discovery of five more bodies, those of Deeming's first wife, Marie (née James), whom he had married in 1881, and their four children, cemented under the floor of a house at Rainhill in England. Under the name of Williams (one of a number of aliases which he used), Deeming, defended by *Deakin, was tried in Melbourne for Emily's murder. He was convicted, sentenced to death, and hanged at the Melbourne Gaol on 23 May 1892. The case aroused a great deal of press and public hysteria and speculation, including allegations that Deeming was 'Jack the Ripper'.

DEMOCRATIC LABOR PARTY (DLP) A minor Australian political party formed after the *Australian Labor Party split during the mid-1950s. The Catholic Social Movement and the Industrial Groups within the Australian Labor Party worked together to oppose communist influence in the trade union movement. *Evatt publicly attacked both groups in 1954, and members were expelled from the Australian Labor Party in 1955. The expelled members formed the Australian Labor Party (Anti-Communist) in 1955, which became the Democratic Labor Party in 1957. The Movement, as it was called, became the National Civic Council in 1957. The Democratic Labor Party was represented

The cartoon character 'Dollar Bill', used to explain **decimal currency** *when it was introduced. (Treasury, Canberra)*

in the Senate until 1974, at times holding the balance of power. A conference of the party in 1978 voted narrowly in favour of its disbandment, but it was soon reconstituted, and continued to field candidates, albeit unsuccessfully, for some years before fading into obscurity.

DENISON, SIR WILLIAM THOMAS (1804–71), governor-general, was born in London, educated at Eton, and graduated from the Royal Military Academy in 1826, after which his work included serving as an engineer officer in Canada and England. He was Lieutenant-Governor of *Van Diemen's Land from 1847 to 1855, during which time he had to resolve the situation involving the *Patriotic Six, was censured following his conflict with the Supreme Court, and took a keen interest in education. Although Denison advocated the continuation of *transportation to the colony, it ended during his term of office. He became *Governor of New South Wales in 1855, and nominal Governor-General of all the Australian colonies. *Responsible government began in New South Wales in 1856. Denison strengthened the defences of New South Wales, attempted to improve the administration of public works, controlled the transfer of Pitcairn Islanders to *Norfolk Island, and shortly before his term ended in January 1861 became involved in controversy with *Cowper over control of the great seal of New South Wales. He was then Governor of Madras from 1861 to 1866. He died in England. His publications include *Varieties of Vice-Regal Life* (1870).

DENNIS, CLARENCE MICHAEL JAMES (1876–1938), journalist and poet, was born to Irish parents at Auburn in South Australia. His father was a hotel-keeper. Dennis was educated at Gladstone Primary School, St Aloysius's College at Sevenhill, and the Christian Brothers' College in Adelaide. He worked first as a clerk, then pursued a journalistic and literary career, interrupted by several brief periods of other employment, including work for the Commonwealth Government during the First World War. He wrote prose and poetry for numerous newspapers and journals, such as the *Bulletin; he and others founded the Adelaide *Gadfly*, which he edited for about eighteen months; and he wrote a daily column for the Melbourne *Herald* from 1922 to 1938. He died in Melbourne. C. J. Dennis is best known for *The Songs of a Sentimental Bloke* (1915). It is written in vernacular verse, and tells the story of the 'Sentimental Bloke', a *larrikin named Bill, and his sweetheart Doreen. Numerous adaptations have been made for film, stage, radio, record and television. His many other publications include *The Moods of Ginger Mick* (1916) (a sequel to *The Songs of a Sentimental Bloke*), *The Glugs of Gosh* (1917), and *Digger Smith* (1918).

DEPRESSION (1840s) Australia's first economic depression. The worst years were 1841–3, and New South Wales and Van Diemen's Land were hardest hit. Insolvencies, unemployment, and poverty were widespread. Several banks failed, including the Bank of Australia, the Sydney Banking Company, and the Port Phillip Bank. Historians have suggested various factors that helped either to cause or to exacerbate the depression: a serious drought in New South Wales during the late 1830s; declining wool prices overseas and wheat prices locally; a diminished flow of English capital to the colonies; too much borrowing on limited security; the withdrawal

DEPRESSION

*Hal Gye's title page for C. J. **Dennis**'s* The Glugs of Gosh *(1917). (Collins Angus & Robertson Publishers)*

of colonial government funds from banks in New South Wales (necessitated largely because land revenue had fallen) to pay for *immigration; and the reduction of commissariat expenditure to New South Wales as a result of the cessation of *transportation to that colony. Most historians stress internal rather than external factors as causes. Various measures were implemented in an attempt to cope with the depression. Assorted legislation in New South Wales introduced more generous provisions for insolvents, abolished imprisonment for debtors and allowed borrowers to use stock and prospective wool-clips as security for advances. Some *squatters resorted to the boiling-down of sheep for tallow. The depression lasted until the mid-1840s.

Depression (1890s) An economic depression experienced by the eastern colonies, Victoria being the most severely affected. The depression lasted from about 1890 to 1894. Many financial institutions collapsed, including building societies, land banks and trading banks. The main bank crashes occurred in 1893. Unemployment was high. *Immigration virtually ceased. Suggested causes, or factors which contributed to the depression, include the decline in overseas prices for wool and wheat; the cessation of overseas capital flow; speculation in real estate, especially in Melbourne and Sydney, during the boom years of the 1880s; overexpansion in both the public and private sectors, especially in the pastoral industry; industrial unrest; and the effects of drought conditions. There is debate as to whether the depression was caused by external or internal factors. Gold discoveries in Western Australia, notably at *Kalgoorlie and Coolgardie, and changes and improvements in Australian primary industries assisted subsequent recovery.

Depression (1930s) *Great Depression

Dictation test *White Australia Policy; *Kisch affair

This cartoon, entitled Ways and Means, *from Australia's first* depression (1840s) *criticized Governor Gipps and his economies. 'Merry' is Francis Mereweather, the immigrant agent in Sydney. (National Library of Australia)*

DIGGER A term with several Australian meanings. During the *gold-rushes of the 1850s, miners became known as diggers. During the First World War, Australian and New Zealand soldiers were called diggers, and the word continued to be used during the Second World War. The word is used also as an informal mode of address among Australian men.

DISCOVERY OF AUSTRALIA Ancestors of the *Aborigines are believed to have settled in Australia more than 40 000 years ago. There is some speculation that the Portuguese were the first Europeans to discover Australia, but the first known European landing was made by *Janssen (or Jansz), a Dutch navigator, in 1606. Numerous Dutch contacts followed, including *Hartog's visit in 1616, the wrecking of the *Batavia* in 1629, and *Tasman's voyages in 1642 and 1644. English contacts began with the wrecking of the *Tryal* off the islands now called the Monte Bello Islands in 1622, the sighting of the coast by the *London* in 1681, *Dampier's visits in 1688 and 1699, and James *Cook's survey of the east coast in 1770. Following the settlement of *New South Wales in 1788, notable British exploration by sea included George Vancouver's survey of part of the south coast in 1791; the voyages of *Bass and *Flinders, together and individually, between 1795 and 1803; James Grant's exploration of parts of the coast of present-day *Victoria in 1800 in 1801; and John Murray's exploration of Western Port, part of King Island, Port Phillip, and Corio Bay in 1801 and 1802. Phillip Parker King, the son of Philip Gidley *King, surveyed a considerable part of the Australian coast between 1817 and 1822. Early French contact included Marion du Fresne's visit to *Van Diemen's Land in 1772; Francois Alesne de St Allouarn's visit to the west coast in 1772; *La Perouse's visit in 1785; Joseph Antoine Raymond Bruny d'Entrecasteaux's visit to Van Diemen's Land and the south coast of the mainland in 1792; and *Baudin's exploration between 1801 and 1803.

DISMISSAL OF THE WHITLAM GOVERNMENT (1975) A political and constitutional crisis. The House of Representatives, which had a Labor majority, and the Senate, which had a non-Labor majority, reached a deadlock following the so-called Loans Affair, which involved attempts by several Labor ministers to raise overseas loans without consulting Cabinet. The Prime Minister, *Whitlam, would not agree to a dissolution, which the Senate was demanding before it would pass the Appropriation Bills (popularly known as Supply). The Governor-General, Sir John Kerr, having obtained advice from the Chief Justice of the *High Court, *Barwick, a former Liberal minister, withdrew Whitlam's commission as Prime Minister, and instead commissioned *Fraser, the Leader of the Opposition, on 11 November. Fraser advised Kerr to dissolve both houses, and the Appropriation Bills were passed. Fraser's Liberal–National Country Party coalition secured majorities in both houses at the elections held on 13 December. The dismissal of the Whitlam Government continues to be the cause of controversy.

DIXON, SIR OWEN (1886–1972), lawyer, was born in Hawthorn, Victoria. His father was a solicitor. He was educated at Hawthorn College and the University of Melbourne, was admitted to the Bar in 1910, and was appointed King's

*Bleak scenes from the **depression** (1890s), published in the* Leader, *14 April 1894. (La Trobe Collection, State Library of Victoria)*

Counsel in 1922. He acted as a Justice of the Supreme Court of Victoria in 1926, became a Justice of the *High Court in 1929, was appointed to the Privy Council in 1957, and was Chief Justice of the High Court from 1952 to 1964. He attempted to establish a consistent interpretation of Section 92 of the *Constitution of the Commonwealth (the section that reads, in part: 'On the imposition of uniform duties of customs, trade, commerce, and intercourse among the States, whether by means of internal carriage or ocean navigation, shall be absolutely free.') During the Second World War he chaired the Central Wool Committee (1940–2), the Australian Shipping Control Board (1941–2), the Marine War Risks Insurance Board (1941–2), and the Allied Consultative Shipping Council of Australia (1942). He was the Australian Minister in Washington from 1942 to 1944. In 1950 he was the United Nations mediator in the Kashmir dispute between India and Pakistan. Sir Owen Dixon died in Hawthorn. His publications include *Jesting Pilate: and other Papers and Addresses* (1965).

DLP *Democratic Labor Party

DOBELL, SIR WILLIAM SMITH (1899–1970), painter, was born in Newcastle, New South Wales. His father was a building contractor. Dobell was an

Dog on the Tucker Box

*William **Dobell**'s* Two Men Carrying a Load *(c. 1943). (National Gallery of Victoria)*

apprentice architect in Newcastle from 1916 onwards. He studied at Julian Ashton's School, Sydney, between 1924 and 1929; at the Slade School, London, in 1929 and 1930; and in Holland in 1930. Dobell remained in Europe and England until 1939, when he returned to Sydney to teach part-time at the East Sydney Technical College. During the Second World War, he served in a camouflage unit and as an official artist in the Civil Construction Corps. Dobell became involved in controversy when his *Portrait of an Artist* (1943), a portrait of Joshua Smith, won the Archibald Prize for 1943. Two artists challenged the validity of the award, on the grounds that the painting was a caricature, not a portrait. Dobell won the subsequent court case. He went on to win two more Archibald Prizes and numerous other awards. He also painted four portraits for covers of *Time* magazine. Dobell, who had become Australia's most famous portrait painter, died at Lake Macquarie. His paintings include *The Cypriot* (1940), *Portrait of a Strapper* (1941), and *Dame Mary Gilmore* (1957).

Dog on the tucker box A bronze statue by an Italian sculptor, F. Rusconi, located on the roadside 8 kilometres (originally 9 miles or 14.5 kilometres before the road was rerouted) from Gundagai, in New South Wales, at a place where teamsters used to camp. The statue, to commemorate the area's pioneers, was unveiled by the Prime Minister, Joe *Lyons, in 1932. It was inspired by an incident involving a bullocky's dog and a tucker box, remembered in a teamsters' ballad called 'Nine Miles from Gundagai', a sanitized version of which was written by Jack Moses, in which the dog sat on the tucker box. The statue has long been a tourist attraction.

DONAHOE (or Donohoe), JOHN (1806?–30), *bushranger, was born in Dublin. Having been sentenced to *transportation for life, he arrived in 'New South Wales in 1825. He and two companions went bushranging in 1827, were captured, and were all sentenced to death. The other two were hanged, but Donahoe escaped. He led a gang of bushrangers in New South Wales until 1830, when he was killed in a fight with police and soldiers in the Bringelly Scrub near Campbelltown. By then, he had become Australia's best-known bushranger. His life and death have inspired many folksongs, including 'Bold Jack Donahoe' and 'The Wild Colonial Boy'.

DORCHESTER LABOURERS *Tolpuddle Martyrs

DOUGLAS CREDIT *Social Credit

DROUGHTS can be described, in general terms, as lengthy periods of rainfall deficiency. The most severe and widespread droughts occurring in Australia since 1860 (when reliable records were established) include those of 1864–6, 1880–6, 1888, 1895–1903, 1911–16, 1918–20, 1939–45, 1964–8, 1976–7, and 1980–3, the worst being that of 1895–1903 (sometimes called the 1902 drought), during which sheep numbers, which had exceeded 106 million in 1892, were halved, cattle numbers fell from more than 12 million in 1895 to 7 million in early 1903, and the average wheat yield for the Commonwealth dropped to 2.4 bushels per acre in 1902.

DRYSDALE, SIR GEORGE RUSSELL (1912–81), painter, was born in England. He settled in Australia with his family in 1923, and was educated at Geelong Grammar School. He later studied art in Melbourne, London and Paris. Russell Drysdale, who spent most of his life in Australia, is probably best known for his outback landscapes. His paintings won numerous awards. Drysdale became a member of the Commonwealth Art Advisory Board in 1962, and a member of the Board of Trustees, Art Gallery of New South Wales, in 1963. His paintings include *The Rabbiter and His Family* (1938), *Sunday Evening* (1941), *The Drover's Wife* (1945), *Two Children* (1946) and *Sofala* (1947).

DUFFY, SIR CHARLES GAVAN (1816–1903), Irish nationalist and Victorian politician, was born in Monaghan, Ireland. His father was a shopkeeper. Duffy was educated at a Presbyterian school. He founded the *Nation*, a weekly journal, in Dublin in 1842, and was admitted to the Bar in Ireland in 1845. As one of the leaders of the Young Irelanders (members of an Irish nationalist movement), Duffy was tried for treason in 1848, but was released after five trials. He was a Member of the British House of Commons from 1852 to 1855. Duffy arrived in Melbourne in 1856, and

DUNTROON

Russell Drysdale's Greek Refugees before Bombed Buildings in Larissa, Greece *(1943). (National Gallery of Victoria)*

worked as a barrister before entering colonial politics. He was a Member of the Victorian Legislative Assembly for several periods (1856–64, 1867–74, 1876–80), was Premier and Chief Secretary in 1871–2, and held various portfolios. He introduced legislation that abolished the property qualification for members of the Legislative Assembly, and was responsible for land legislation in 1862 that was designed to help *selectors, but which failed, and helped *squatters more than selectors. Duffy, a liberal, also supported *protection and federation. He was one of the founders of the *Advocate*, a Catholic journal, which began in 1868. Duffy lived in Nice, France, from 1880 until his death. His publications include *Young Ireland* (1880), *Four Years of Irish History* (1883), *The League of North and South* (1886), *Conversations with Carlyle* (1892), and *My Life In Two Hemispheres* (1898).

DUNTROON *Army; *Bridges, Sir William Throsby; *Campbell, Robert

DYSON FAMILY A well-known Australian artistic and literary family. Its best-known member is WILLIAM (WILL) HENRY DYSON (1880–1938), artist and writer, who was born at Alfredton, near Ballarat. His father worked on the goldfields as a miner and hawker. The family moved to Melbourne when Dyson was a child, and he was educated at Albert Park State School. Dyson drew cartoons for the *Bulletin* and other journals, became friendly with the *Lindsay family, and played a leading role in Melbourne's bohemian society. He married Ruby Lindsay in 1909, and shortly after their marriage they sailed for England. Dyson was an extremely successful radical cartoonist on the London *Daily Herald* from 1912 to 1916. He then became an Australian war artist, and was wounded twice. He again worked on the *Daily Herald* from 1918 to 1921, but was greatly affected by his wife's death, in 1919, during the *'Spanish' influenza pandemic. Dyson did not regain his pre-war force as a cartoonist, although his most famous cartoon, *Peace and Future Cannon Fodder*, was published in 1919. In 1925 he returned to Australia, where he drew for the Melbourne *Herald*. He also drew for *Punch*, and later for *Table Talk*. He took up etching, which later brought him further fame in England and the USA. He returned to England, via America, in 1930 and worked again as a cartoonist on the *Daily Herald* from 1931 until his death. Dyson's publications include *Cartoons* (1914), *Kultur Cartoons* (1915), *Will Dyson's War Cartoons* (1916), *Australia at War* (1918), and *Artist Among the Bankers* (1933). Other well-known members of the family include **AMBROSE (AMB) ARTHUR DYSON** (1876–1913), Will's elder brother, cartoonist and illustrator, whose subjects include *larrikins; **EDWARD AMBROSE (AMBY) DYSON** (1908–53), Ambrose's son, cartoonist; and **EDWARD GEORGE DYSON** (1856–1931), Will's eldest brother, writer, whose publications include *Rhymes from the Mines* (1896), *Below and On Top* (1898), *The Gold-stealers* (1901), *In the Roaring Fifties* (1906), and *Spats' Fact'ry* (1914).

*Will **Dyson**'s lithograph* The Wine of Victory—German Prisoners—The Salient *captures some of the tragedy of the First World War. (Australian War Memorial)*

E

EARDLEY-WILMOT, SIR JOHN EARDLEY (1783–1847), colonial administrator, was born in London and educated at Harrow. His father was a master in chancery. Eardley-Wilmot, who was created a baronet in 1821, was called to the Bar in 1806, chaired the Warwickshire Quarter Sessions from 1830 to 1843, and was a Member of the House of Commons from 1832 to 1843. During his term of office as Lieutenant-Governor of *Van Diemen's Land, which began in 1843, he was faced with the *depression of the 1840s; had to implement the unpopular new probation system; was involved in various conflicts with the Anglican Bishop Nixon; and was a central figure in the constitutional crisis involving the *Patriotic Six. Eardley-Wilmot was recalled in 1846, allegedly for neglecting the convict system. He was informed privately that he would not be employed in an official position again because of alleged rumours about his personal life. He died in Van Diemen's Land the following year. His publications include *An Abridgement of Blackstone's Commentaries* (1822) and *A Letter to the Magistrates of England* (1827).

ECCLES, SIR JOHN CAREW (1903–), neurophysiologist, was born in Melbourne and educated at Warrnambool and Melbourne High Schools and the University of Melbourne, before going to Oxford University as the Victorian Rhodes Scholar in 1925. He gained his D Phil at Oxford, where he remained as a Fellow at Exeter College from 1927 to 1934 and as a Fellow at Magdalen College and Lecturer in Physiology from 1934 to 1937. He returned to Australia as Director of the Kanematsu Memorial Institute of Pathology at Sydney Hospital from 1937 to 1943. After a period as Professor of Physiology at the University of Otago in New Zealand from 1943 to 1951, he again returned to Australia, this time to become Professor of Physiology at the John Curtin School of Medical Research at the Australian National University in 1951, a position which he held until 1966. From 1957 to 1961, Eccles, who was knighted in 1958, was the President of the Australian Academy of Science. In October 1963 he was awarded the Nobel Prize for Physiology and Medicine (with A. L. Hodgkin and A. F. Huxley) for his work on the chemical processes of nerve cell impulses. His many other awards and honours include the Royal Society's Royal Medal, which he received in 1962. He went on to become a professorial lecturer at the Institute of Biomedical Science in Chicago from 1966 to 1968 and Distinguished Professor at the State University of New York at Buffalo from 1968 to 1975. His numerous publications include *Reflex Activity of the Spinal Cord* (jointly) (1932), *Neurophysiological Basis of Mind* (1953), *Physiology of Nerve Cells* (1957), *Facing Reality: Philosophical Adventures of a Brain Scientist* (1970), and *The Self and Its Brain* (jointly) (1977).

EIGHT-HOUR DAY MOVEMENT Nineteenth-century campaign for shorter working hours. Unionists campaigned for the reduction of the working week (which then comprised six days) from sixty to forty-eight hours. Sydney stonemasons and Melbourne building workers, the latter having formed the Eight Hours League, became the first to gain the reduction in 1856. Annual celebrations were introduced in Melbourne and later elsewhere. Many unions were first formed for the purpose of fighting for the eight-hour day. Most Australian

OPPOSITE PAGE
At annual marches commemorating the **Eight-Hour Day Movement** *workers have traditionally carried trade union banners such as these ones, published in the* Australasian Sketcher, *14 June 1873. (La Trobe Collection, State Library of Victoria)*

workers had gained it by the end of the nineteenth century. All of the states, the Australian Capital Territory, and the Northern Territory now commemorate the movement with an annual public holiday, held on various dates and known by various names (such as Labour Day and Eight Hour Day).

EMANCIPISTS   A term used during the period of *transportation. It had several different meanings. In a narrow sense, it referred only to those convicts who had been pardoned, conditionally or absolutely, by the governor. In a broader sense, it was applied to all ex-convicts, and this usage was more common. The term was also applied to members of a political group, consisting of emancipists and liberals, that campaigned for legal and political reforms in New South Wales during the 1820s, 1830s, and early 1840s. *Wentworth was for many years the acknowledged leader of this group, which in 1835 founded the *Australian Patriotic Association. Emancipists and *exclusives came into conflict over various issues relating to the former's position in colonial society.

EQUAL PAY   The right of women to receive the same wages and salaries as men. The principle of equal pay was introduced in Australia in stages. Queensland legislated for equal pay as early as 1916, but the legislation proved to be ineffective. In 1958 New South Wales legislated for 'equal pay for equal work' to be phased in over five years. This legislation differentiated between men's and women's work. Women performing 'men's work' were to receive equal pay; women performing 'women's work' (such as typing) were not. Other states had introduced similar legislation by the late 1960s. The Commonwealth Conciliation and Arbitration Commission accepted the principle of 'equal pay for equal work', to be phased in over three years, in 1969. Equal pay was not to be provided 'where the work in question is essentially or usually performed by females'. In 1972 the Commission accepted the principle of 'equal pay for work of equal value', to be gradually introduced by mid-1975, but rejected a claim for the introduction of an equal minimum wage for men and women. The Commission introduced the principle of the same minimum wage for men and women in 1974. Acceptance of equal pay in theory has not led to equal incomes in practice.

ERN MALLEY HOAX   A significant Australian literary *cause célèbre* of the 1940s. Two soldiers, James McAuley (1917–76) and Harold Stewart (1916–), both of whom were conservative and traditional poets, executed the hoax as an experiment to see whether those who wrote and admired modernist poetry could distinguish between real poetry and nonsense. Using several dictionaries, a collection of Shakespeare's plays, and an American report on swamp drainage, they concocted a series of sixteen poems entitled 'The Darkening Ecliptic', purportedly written by Ern Malley, a twenty-five-year-old Sydney poet, motor mechanic, and insurance salesman who had died in 1943. Claiming to be Ethel Malley, Ern's sister (who was just as fictitious as her brother), they sent the poems to Max Harris, the leader of the modernist *Angry Penguins, who praised them and published them in the autumn 1944 issue of the *Angry Penguins* journal. Following the exposure of the hoax in the Sydney *Sun* on 5 June 1944, a spirited debate took place between conservatives and modernists, some of

whom continued to see merit in the poems. Harris was prosecuted in South Australia for publishing 'indecent advertisements' within the poems, was found guilty, and was fined £5. The hoax, which helped to bring about the demise of the journal, is considered to have been a setback for the modernist movement in Australia.

ESCAPE CLIFFS *Northern Territory; *Port Essington

EUREKA FLAG Flag flown by *diggers during the *Eureka Rebellion, and later by other groups, including striking shearers at *Barcaldine and, more recently, members of the Builders' Labourers' Federation. The Flag has a white cross and five white stars, one in the centre and one at each of the four ends of the cross, representing the constellation of the Southern Cross, on a blue background.

EUREKA REBELLION An armed conflict in 1854 between *diggers and government authorities on the gold-fields at Ballarat, Victoria. The diggers' grievances included the licence system and its administration (under which diggers were forced to pay thirty shillings a month, had to carry their licences at all times, and had to face licence hunts, which were deeply resented), corruption among officials, lack of political representation, and limited access to land. A digger named Scobie was kicked to death near the Eureka Hotel on 7 October. Bentley, the hotel licensee, and three others were charged with Scobie's murder. Magistrates discharged them on 12 October, strengthening the diggers' suspicions of corruption. Diggers held a mass protest meeting on 17 October, after which the Eureka Hotel was burnt down. As a result, three diggers were charged on 21 October with riot. At another mass meeting on 11 November, diggers formed the Ballarat Reform League. Bentley was eventually convicted of Scobie's murder on 23 November. Ballarat Reform League representatives demanded the release of the three charged with riot. *Hotham, the Lieutenant-Governor (later Governor) of Victoria, refused, and sent more troops to Ballarat. When soldiers entered Ballarat on 28 November they scuffled with diggers, and next day at a mass meeting where the Southern Cross flag (now usually called the *Eureka flag) was displayed, diggers resolved to burn their licences. The authorities conducted another licence hunt the following day. About 1000 diggers, led by *Lalor, built a stockade ('breastwork or enclosure of upright stakes') on the Eureka lead. About 280 soldiers and police attacked the stockade on 3 December, rapidly overpowering the 150 or so diggers inside at the time. They took more than 100 prisoners. About thirty diggers and five soldiers and police were killed. Martial law was proclaimed and remained in force for several days. Thirteen diggers were charged with high treason; the prosecution against one was dropped and juries acquitted the others. Rewards were offered for information leading to the capture of Lalor and two others, but were later withdrawn. A royal commission was appointed in 1854 to inquire into the gold-fields. Its recommendations, most of which were adopted, included the abolition of the licence fee, the introduction of an export duty on gold, and the introduction of a miner's right, which conferred legal and political rights, to cost £1 a year.

EVATT, HERBERT VERE (1894–1965), lawyer and politician, was born at Maitland, New South Wales. His father was a publican. He was educated at Fort Street High School and the University of Sydney, admitted to the Bar in 1918, and appointed King's Counsel in 1929. Evatt was a Justice of the *High Court from 1930 to 1940, and Chief Justice of New South Wales from 1960 to 1962. He was a Member of the Legislative Assembly of New South Wales from 1925 to 1930, serving as a Labor Member to 1927, then as an Independent Labor Member. Evatt was a Labor Member of the House of Representatives from 1940 to 1960, and held various offices, including those of Attorney-General (1941–9) and Leader of the Opposition (1951–60). The split in the Labor Party that resulted in the formation of the *Democratic Labor Party occurred during the period of Evatt's leadership of the federal parliamentary party. He was prominent in the early development of the *United Nations Organization; he presided over the United Nations General Assembly (1948–9), and chaired the United Nations Atomic Energy Commission (1947) and the United Nations Palestine Commission (1947). Evatt, who was a Doctor of Laws and a Doctor of Letters, wrote a number of books including *The King and His Dominion Governors* (1936), *Rum Rebellion* (1938), and *Australian Labour Leader: W. A. Holman and the Labor Movement* (1940).

EXCLUSIVES Sometimes known as exclusionists. The name applied in New South Wales during the *transportation period (especially the 1820s and 1830s) to those people who opposed the restoration of complete civil rights to *emancipists.

EXILES Convicts who had served a prison term in England, and in some cases had laboured on public works, before undergoing *transportation to Australia, where they were given conditional pardons or *tickets of leave. A large proportion of the exiles were sent to the *Port Phillip District.

EXPLORATION OF AUSTRALIA by land by Europeans mostly took place in the nineteenth century. Early exploration in eastern Australia included the crossing of the *Blue Mountains, John Oxley's expedition to Port Macquarie in 1818, the *Hume and Hovell expedition to the south in the mid-1820s, and Allan Cunningham's discovery of the Darling Downs in 1827. *Sturt solved the 'riddle of the rivers' in the 1820s and 1830s, and also explored in the centre of Australia in the mid-1840s. *Mitchell explored in the south-east in the 1830s, and in the north in the mid-1840s. Paul Edmund de Strzelecki discovered and named Mount Kosciusko in 1840, and then crossed Gippsland (which Angus McMillan had already explored). *Grey undertook expeditions in Western Australia in the 1830s. *Eyre explored parts of South Australia from 1839 onwards, and became the first European to cross the desert now known as the Nullarbor Plain. *Leichhardt explored in the north-east and north in the 1840s, finally disappearing during an expedition. Edmund Kennedy led a disastrous expedition in the north-east in 1848. Augustus Charles Gregory explored in the west in the 1840s, in the north and north-east in the 1850s, and from central Queensland to Adelaide in 1858. In 1861, some members of the *Burke and Wills expedition crossed the continent from south to north, but

EYRE, EDWARD JOHN

*Portrait of the explorer Edward John **Eyre**, the first White person to cross the Nullarbor Plain. (Mortlock Library of South Australiana, State Library of South Australia)*

died on the way back. *Stuart, after two unsuccessful attempts in 1860 and 1861, also crossed from south to north in 1862. John *Forrest explored in Western Australia and South Australia in the 1860s, 1870s, and 1880s; in 1870 he crossed the Nullarbor Plain area from west to east. *Warburton's party became the first to cross Australia from the centre to the west (1873–4). Ernest Giles explored in the centre of the continent in the 1870s and crossed much of the continent from east to west and back again in 1875 and 1876. Many other explorers and *overlanders also opened up parts of the continent. Much of Van Diemen's Land was explored in the first few decades of white settlement.

EYRE, EDWARD JOHN (1815–1901), explorer, was born in England. His father was a vicar. Eyre migrated to New South Wales in 1833, became a *squatter, and later overlanded stock to the *Port Phillip District and South Australia. After moving to Adelaide, he undertook expeditions to the Flinders Ranges and the Eyre Peninsula in 1839. He took stock by sea to *Albany and overlanded them to Perth in 1840. Later that year he made unsuccessful attempts to find a route from Adelaide to the west. However, in 1841, he became the first white person to cross the area now known as the Nullarbor Plain. Eyre, John Baxter, and three Aborigines left Fowler's Bay in February; in April two of the Aborigines killed Baxter and fled; Eyre and the other Aborigine, Wylie, having been helped by French *whalers near present-day Esperance, finally reached Albany in July. Eyre was the resident magistrate and protector of Aborigines at Moorundie in South Australia from 1841 to 1844. In 1846 he was appointed Lieutenant-Governor of New Zealand, a position he held until 1853. In 1854 he was appointed Lieutenant-Governor of St Vincent (in the West Indies), held that position until 1860, was acting Governor of the Leeward Islands for some time, and became acting Governor of Jamaica in 1864. His controversial handling of a Negro uprising there in 1865 led to his recall. Eyre died in England. He published *Journals of Expeditions of Discovery into Central Australia and Overland from Adelaide to King George's Sound in the Years 1840–1* (1845).

F

FADDEN, SIR ARTHUR WILLIAM (1895–1973), was born in Queensland, educated at Walkerston State School, and was assistant Town Clerk (1913–18) and Town Clerk (1918) at Mackay before practising as a chartered accountant in Townsville. Fadden was a Country Party member of the Queensland Legislative Assembly from 1932 to 1935, then a Country Party Member of the House of Representatives from 1936 to 1958, becoming Leader of the Country Party in 1941. He held various portfolios. After *Menzies resigned as Prime Minister, Fadden replaced him, leading a *United Australia Party–Country Party Government. His prime ministership was short-lived (29 August–7 October 1941). Labor and the Independents defeated Fadden's budget, and *Curtin became the next Prime Minister. Fadden was Leader of the Opposition until 1943, and continued to lead the Country Party until his retirement from Parliament in 1958. From 1949 to 1958 he was Deputy Prime Minister and Treasurer in the Menzies Government. He pursued business interests after leaving politics.

FAIRFAX, JOHN (1804–77), newspaper proprietor, was born in Warwick, England. In 1836, while running a newspaper in England, he went bankrupt, but later paid his creditors. Fairfax migrated to Sydney in 1838, and became the Librarian of the Australian Subscription Library in 1839. He and Charles Kemp bought the daily *Sydney Herald* in 1841, a paper that had been founded as a weekly in 1831. Its name was changed to *Sydney Morning Herald* in 1842. Fairfax bought Kemp's interest in 1853, and his sons joined him in the business. The *Sydney Morning Herald* became the major daily newspaper in New South Wales. Fairfax also became a foundation director of the Australian Mutual Provident Society, a director of the Sydney Insurance Company and various other institutions, a trustee of the Savings Bank of New South Wales, and president of the Young Men's Christian Association. He was actively involved in the Congregational Church, and was nominated to the Legislative Council in 1874. His sons continued to run the newspaper after his death in Sydney.

FARRER, WILLIAM JAMES (1845–1906), wheat-breeder, was born in England, the son of a tenant-farmer. Farrer was educated at Christ's Hospital, London, and Pembroke College, Cambridge. Ill health forced him to abandon his medical studies. In 1870 he migrated to Australia, where he worked as a tutor on a sheep station. Farrer qualified as a surveyor in 1875 and worked in that capacity with the Department of Lands in New South Wales from 1875 to 1886. He then settled on the Murrumbidgee River, near present-day Canberra, and experimented with wheat-breeding from 1886 to 1898, when he was appointed wheat experimentalist with the Department of Agriculture, a position he held until his death. Farrer's work contributed greatly to the development, in Australia and other countries, of the wheat industry. His best-known wheat, 'Federation' (named in 1901), was the most widely grown wheat in Australia from 1910 to 1925.

FAWKNER, JOHN PASCOE *Melbourne; *Strutt, William

FEDERAL CONVENTION (1897–8) A significant event in the *federation movement leading to the creation of the *Commonwealth of Australia. The Convention, which had been proposed at the Corowa Conference, had as its

FEDERATION MOVEMENT

*Mick Armstrong's "Hulloa, Caretaker!". Arthur **Fadden** was Prime Minister of Australia from 29 August to 7 October 1941. (La Trobe Collection, State Library of Victoria)*

purpose the drafting of a federal constitution to be put to the people at a referendum. It comprised ten delegates each from New South Wales, Victoria, South Australia, Tasmania, and Western Australia (the delegates being elected by voters in each colony, except in Western Australia, where they were elected by members of parliament). Queensland did not send delegates. Notable delegates included *Barton, *Deakin, *Kingston, and *Reid. The Convention met in Adelaide and Sydney in 1897, and in Melbourne in 1898. A federal constitution was drafted. The first federal referendum (held in four colonies in 1898) failed; after amendments were made, the second (held in five colonies in 1899, and one in 1900), passed in all of the Australian colonies.

FEDERAL COURT OF AUSTRALIA *Commonwealth Conciliation and Arbitration Court

FEDERATION CUP *Tennis

FEDERATION MOVEMENT The movement that led to the federation of the Australian colonies in 1901. Politicians and others had proposed federation as early as the 1840s, but serious attempts were not made to introduce it until the 1880s and 1890s. The Federal Council, which was formed in 1885 and met regularly for the rest of the nineteenth century, was weakened by the absences of South Australia (except for two years) and New South Wales. *Parkes publicly supported federation in his *Tenterfield Oration in 1889. Those in favour of federation argued that uniform policies on matters such as defence, *immigration, and tariffs would be advantageous. Federation was discussed at various conferences, including the Australasian Federation Conference, held in Melbourne in 1890; the National Australasian Convention, held in Sydney in 1891, at which a draft constitution was produced; the so-called Corowa Conference in 1893, attended by representatives of the Australasian Federation League, the *Australian Natives' Association, and other organizations; the Premiers' Conference held in Hobart in 1895; the *Federal Convention, held in Adelaide and Sydney in 1897 and Melbourne in 1898, at which another draft constitution was produced; and the Premiers' Conference held in Melbourne in 1899. Two federal referendums were held, based on the draft constitution produced at the Federal Convention. The first was held in New South Wales, Victoria, Tasmania, and South Australia in 1898. 'Yes' majorities were recorded in each of these colonies, but the 'Yes' majority in New South Wales was insufficient to carry the referendum there. The draft constitution was amended and a second referendum was held in those colonies in 1899. Increased 'Yes' majorities were recorded in each of them. A 'Yes' majority was also recorded in Queensland, when the referendum was held there later in 1899. Western Australia voted in favour of federation at the referendum held there in 1900, shortly after the British Parliament had passed the *Commonwealth of Australia Constitution Act*. Federation occurred on 1 January 1901. The six colonies became states of the *Commonwealth of Australia, which was part of the British Empire. Fiji and New Zealand had shown interest in the early stages of the movement, but ultimately did not join the Australian colonies in federation.

'FEDERATION' WHEAT *Farrer, William James

FELTON, ALFRED (1831–1904), businessman and philanthropist, was born in England, where he is believed to have been apprenticed to a chemist. His father was a tanner. Felton migrated to Victoria in 1853, transported goods during the *gold-rushes, and then became a merchant, and later a pharmaceutical chemist, in Melbourne. He and Frederick Sheppard Grimwade bought a drug-house, which they renamed Felton, Grimwade & Co., in 1867; it became the largest in Victoria, with some interests elsewhere. They also founded various other businesses during the 1870s and 1880s, and Felton bought two large rural properties in partnership with Charles Campbell. Felton left money to establish a trust fund, originally of approximately £380 000: half of the income was to be used for charities, especially those helping women and children; the other half to buy art works for the Melbourne National Gallery. The Felton Bequest has been, and continues to be, of great importance to the gallery.

FEMALE FACTORY A name that usually refers to an institution that existed during the *transportation period at *Parramatta, New South Wales. *King established it in 1804. The Female Factory provided asylum, punishment, and sometimes employment (such as cloth-making), for convict and other women. A new building was erected near Parramatta in 1821, extended in 1828, and again in 1839. It housed female convicts who had been sent there on arrival, who were unable to be assigned, or who had been returned from assignment. It also housed convict and other women who had committed crimes within the colony. The largest numbers housed there were 1203 women and 263 children in 1842. It was closed in 1848. There was a similar institution at the Cascades, near Hobart.

FINANCIAL AGREEMENT (1927) An agreement between the Commonwealth and state governments. The Australian people approved it at a referendum in 1928, and the *Constitution of the Commonwealth of Australia was amended accordingly in 1929. A Loan Council, with the authority to raise loans for all Australian governments, except for defence purposes, was established. The Financial Agreement helped to transfer power from the states to the Commonwealth.

FIRST FLEET The eleven ships that carried the first White settlers (convicts, marines, and civilians) to Australia. It sailed from England in 1787 under the command of *Phillip. The ships were HMS *Sirius*, HMS *Supply*, the *Alexander*, the *Charlotte*, the *Friendship*, the *Lady Penrhyn*, the *Prince of Wales*, the *Scarborough*, the *Borrowdale*, the *Fishburn*, and the *Golden Grove*. The Fleet arrived at *Botany Bay, but the site was considered unsuitable, so on 26 January 1788 it landed at Sydney Cove in Port Jackson. One thousand and thirty people went ashore; of these 548 men and 188 women were convicts. The colony of New South Wales was formally proclaimed on 7 February.

FIRST WORLD WAR (1914–18) A major war, also known as the Great War, fought between the Allied and Associated Powers (including Serbia, Britain, Belgium, Russia, France, Japan, Canada, New Zealand, Australia, Italy (from 1915), Rumania (from 1916), the USA (from 1917), Montenegro, Greece, and Portugal) and the Central Powers (including Germany, Austria-Hungary, Turkey (from 1914), and Bulgaria (from 1915)). The long-term causes, complex and numerous, included international tensions over naval, trade, and colonial matters, exacerbated by the existence of various international alliances. The immediate cause was the assassination of Archduke Franz Ferdinand, the Austro-Hungarian heir, by a Serbian nationalist at Sarajevo in June 1914. Fighting began in August. The main areas of fighting during the war were on the Western, Eastern, Balkan, and Italian Fronts, and in Palestine, Mesopotamia, and Africa. The main naval operations occurred in the Pacific and South Atlantic Oceans, and the Adriatic and North Seas. The Central Powers were defeated. The armistice with Germany, which marked the end of fighting, was signed on 11 November 1918. After the war the provisions of various treaties, including the Treaty of *Versailles, were determined at the *Paris Peace Conference. An estimated ten million people died directly because of the First World War. The *'Spanish' influenza pandemic, which began in the war's latter stages, is believed to have claimed an even greater number of lives. Australia, as a member of the British Empire, automatically followed Britain to war in August 1914. The *Australian Imperial Force, raised by *Bridges, was Australia's main expeditionary force during the war. An Australian Navy and Military Expeditionary Force was also raised. The *Royal Australian Navy and the Australian Flying Corps (the predecessor of the *Royal Australian Air Force) took part. All members of the Australian forces who served overseas were volunteers. At the beginning of the war, Australian forces quickly captured German *New Guinea. During the war, the main areas in which they fought were the Dardanelles, during the *Gallipoli campaign, which led to the creation of the *Anzac legend; France and Belgium,

First World War

Much of the huge amount of propaganda published in Australia during the **First World War**, *such as this poster issued by the Queensland Recruiting Committee, was directed at women. Why the treatment of Queensland women would be worse than that of French and Belgian women and children is unclear. (Australian War Memorial)*

where Australian soldiers were particularly prominent in the attacks at Pozieres (July–September 1916), Bullecourt (April–May 1917), Third Ypres (September–November 1917), the counter-attack at Villers-Bretonneux (April 1918), and the assault at Amiens (August 1918); and Egypt and Palestine, where the Australian Light Horse played a notable role. They were also involved in minor engagements elsewhere. Prominent Australian soldiers included *Chauvel, *Gellibrand, and *Monash. The Australian forces, whose members were often used as shock troops, suffered extremely high casualty rates. Australia's population in 1916 was 4 875 325. During the war, the Australian troops raised (excluding naval forces) numbered 416 809; of these 331 789 took the field. The battle casualties numbered 215 045; of these, 59 342 were killed or died of wounds. There were also naval casualties, including 170 deaths. It should be noted that many of the other countries involved in the war relied upon *conscription. Two attempts to introduce conscription for overseas service, led by *Hughes and opposed by people such as *Mannix, failed. The conscription issue was a major factor leading to the split that occurred in the *Australian Labor Party during the war. The Commonwealth Government, whose powers increased during the war, introduced wartime legislation such as the *War Precautions Act* 1914. The *Industrial Workers of the World became the subject of controversy. The Returned Sailors' and Soldiers' Imperial League of Australia, later the *Returned Services League of Australia, was founded during the war years.

FISHER, ANDREW (1862–1928), politician, was born in Scotland and educated at a local school. He began to work in mines when he was about ten, and became involved in the Ayrshire Miners' Union. He migrated to Queensland in 1885 and held various jobs in mining areas, including those of manager and engine-driver, and his union involvement continued: he became secretary and president of the Gympie branch of the Amalgamated Miners' Association. He was also involved in the newly formed Labor Party; he was president of the Gympie branch, and was elected to the Queensland Legislative Assembly in 1893. After losing his seat in 1896, Fisher returned to engine-driving, acted as auditor for the municipal council, and helped other Labor supporters to establish the Gympie *Truth*. He was re-elected in 1899 (without formal Labor endorsement), and was Secretary for Railways and Public Works in the *Dawson Government, the first Labor Government in the Australian colonies, which lasted for only a few days in

During Matthew Flinders's circumnavigation of Australia from 1801 to 1803, William Westall painted this watercolour entitled A View of King George's Sound *(1802). (National Gallery of Victoria)*

OPPOSITE PAGE

TOP RIGHT
C. Dudley Wood's water-colour, Seagulls over Eastern Coastline (1938), for the cover of Australia To-day, 'The Commonwealth's Pictorial National Annual', published by the United Commercial Traveller's Association. The **Commonwealth of Australia** is the official title of the Australian nation. (The University of Melbourne Archives)

BOTTOM RIGHT
Travel poster by Moody advertising **South Australia**'s centenary celebrations of 1936. (La Trobe Collection, State Library of Victoria)

LEFT
Travel poster by James Northfield showing some of the attractions of **Canberra**. (La Trobe Collection, State Library of Victoria)

December 1899. Fisher campaigned for federation, and was a Member of the House of Representatives from 1901 to 1915. He was Minister for Trade and Customs in the *Watson Labor Government (April–August 1904). He became Labor Leader on Watson's resignation in 1907, and Prime Minister and Treasurer in a minority Labor Government in November 1908. His Government was defeated by the newly formed *Fusion Party in May 1909. Fisher was again Prime Minister and Treasurer following Labor's win in the 1910 elections; this Government's achievements included the provision of maternity allowances, the extension of old-age and invalid pensions, and the establishment of the Commonwealth Bank (later the *Commonwealth Banking Corporation). Following Labor's loss in the 1913 elections, Fisher became Leader of the Opposition. The Liberals controlled the House of Representatives, but not the Senate, and there was a double dissolution in 1914. Labor won the elections, and Fisher again became Prime Minister and Treasurer. He resigned in 1915 and served as Australian High Commissioner in London from 1916 to 1921. He died in England.

FISHER'S GHOST is the central figure in Australia's best-known ghost story. There are many different accounts of the story. Frederick George James *Fisher, a *ticket-of-leave convict who farmed at Campbelltown, about 50 kilometres south-west of *Sydney, disappeared in 1826. George Worrall, another ticket-of-leave convict who farmed an adjoining property, claimed that Fisher had left Australia to avoid a forgery charge and began to take his assets, arousing some suspicion. Weeks (or months, according to other versions) later, a man named John Farley allegedly saw the ghost of Fisher on or near a fence at Fisher's property. A reward for the discovery of Fisher's body was offered, bloodstains were found on the fence where Farley had 'seen' the ghost, and Fisher's body was found near by. Worrall was tried for the murder, found guilty, and hanged in 1827.

FLAGS *Aboriginal flag; *Australian national flag; *Eureka flag

FLINDERS, MATTHEW (1774–1814), navigator, was born in England. His father was a surgeon. In 1789 Flinders joined the navy; he sailed to the South Seas with *Bligh (1791–3), took part in the battle of the 'Glorious First of June', a British naval victory against the French in 1794, and sailed to New South Wales in HMS *Reliance* in 1795. He and *Bass explored parts of *Botany Bay in 1795. After visiting *Norfolk Island in the *Reliance*, he and Bass explored from Port Jackson to present-day Lake Illawarra in 1796. Both sailed to Cape Town in the *Reliance* (1796–7). Flinders (as lieutenant) sailed in the *Francis* to the *Furneaux Islands in 1798. Convinced of the existence of a strait between New South Wales and *Van Diemen's Land, Flinders (in command) and Bass sailed south from Port Jackson in the *Norfolk*, confirmed the existence of the strait (now known as Bass Strait), discovered Port Dalrymple, and circumnavigated Van Diemen's Land (1798–9). Flinders then sailed the *Norfolk* north from Port Jackson to near present-day Bundaberg, charting part of the coast. He returned to England in the *Reliance* in 1800. He was promoted commander and placed in charge of HMS *Investi-

*A romanticized version of the discoveries of James **Cook** and Jean-François de Galaup, Comte de La Pérouse, is presented in Antoine Phelippeaux's Tableau des decouvertes du Capne. Cook & de la Perouse. (National Library of Australia)*

*Aboriginal spokesman and artist William **Barak** did much to record his people's culture. The European-style clothing worn by some of the figures in his Corroboree indicates that this corroboree or dance ceremony occurred after White settlement. (National Gallery of Victoria)*

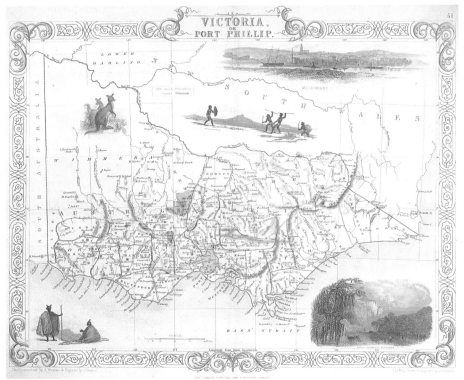

*Viennese-born Eugen von **Guérard**'s painting* Tower Hill *(1855), commissioned for Western District settler James Dawson, is considered to be a good guide to the flora in the Tower Hill area, near Warrnambool in Victoria, in the days before White settlement. (Warrnambool Art Gallery)*

*Map of **Victoria** or the Port Phillip District (c. 1851), drawn and engraved by J. Rapkin and published by John Tallis & Co., London and New York.*

gator, with instructions to survey the coast of *New Holland. Despite the *Investigator*'s poor condition, Flinders circumnavigated the continent between 1801 and 1803, charting large sections of the coast. He met *Baudin at Encounter Bay in early April 1802. Shortly afterwards he entered Port Phillip, and climbed the highest peak (later named Flinders Peak) in the You Yangs. Sailing to England to request the use of another vessel, Flinders was aboard HMS *Porpoise* when it was wrecked in 1803. He navigated the cutter back to Port Jackson, a distance of some 1200 kilometres, to seek help. He sailed for England once again that year, this time in charge of the schooner *Cumberland*. As the schooner was leaking badly, he sought aid in Mauritius, but Britain and France were at war, and the French confined him on Mauritius until 1810. He then returned to England, where he was promoted post-captain. He died in London. His publications include *Observations on the Coasts of Van Diemen's Land, on Bass's Strait and its Islands, and on part of the Coasts of New South Wales* (1801) and *A Voyage to Terra Australis* (1814).

FLINDERS ISLAND *Furneaux Islands; *Robinson, George Augustus

FLOREY, HOWARD WALTER, BARON OF ADELAIDE AND MARSTON (1898–1968), pathologist, was born in Adelaide, and educated at St Peter's College and the University of Adelaide, before taking up a Rhodes Scholarship at Oxford University in 1921. He also studied in the USA. Florey was Professor of Pathology at Sheffield University from 1931 to 1935 and at Oxford University from 1935 to 1962. He and Dr Ernst Chain developed penicillin, the first antibiotic, enabling it to be manufactured on a large scale during the latter part of the Second World War. For this work they were awarded the Nobel Prize for Physiology and Medicine in December 1945, with Sir Alexander Fleming, who had discovered penicillin in 1928. Florey was an adviser to the John Curtin School of Medical Research in Canberra from 1947 to 1957; became Provost of Queen's College, Oxford, in 1962, and was installed as Chancellor of the Australian National University in 1966. Knighted in 1944, Florey became the first Australian president of the Royal Society in 1960, was appointed a life peer in 1965, and received many other honours and awards. He died in Oxford. He had published many scientific papers, co-written *Antibiotics* (1949), and edited *General Pathology* in 1954, 1957, and 1962.

FLYING DOCTOR SERVICE *Royal Flying Doctor Service

FLYNN, JOHN *Royal Flying Doctor Service

FOOTBALL has been played throughout Australia for many years. Rugby Union (an amateur game), Rugby League (a professional game), and Soccer have been particularly strong in New South Wales and Queensland; Australian Rules in Victoria, South Australia, Western Australia and Tasmania. The game that became known as Rugby Union was first played in Australia in 1864, Rugby League in 1907, and Soccer in 1880. Australian Rules was first played in the 1850s. The latter is a fast, free-flowing sport, characterized by aerial contests, physical contact, and long kicking. It is

OPPOSITE PAGE
Illuminated address presented to Charles Alfred Topp by the Premier of Victoria, 1899, for his contribution to the **federation movement**. *(The University of Melbourne Archives)*

Football

Football scenes from the 1928 Victorian Football League grand final, in which Collingwood defeated Richmond, published in the Australasian, *6 October 1928. (La Trobe Collection, State Library of Victoria)*

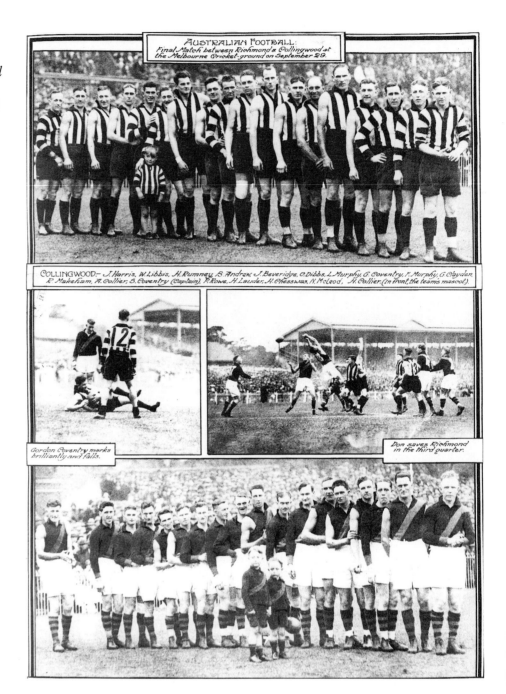

played on an oval field with an oval ball. Each side now has twenty players, including two interchange players. A game is played over four quarters, each twenty-five minutes long, with time added to compensate for delays in play. The Melbourne Football Club was formed in 1858; soon after, several other Victorian clubs were formed, and the rules of the new game evolved over the next few years. Henry Colden Antill Harrison, the 'father of Australian Rules football', drafted a code that was adopted in 1866. The game spread to the other colonies (and briefly to New Zealand and elsewhere). Developments included the introduction of umpires in 1872, uniforms for players in 1873, 'behind' posts on either side of the goal posts (a 'behind' being worth one point, a goal six points) in 1897, and boundary umpires in 1903. Eight clubs left the Victorian Football Association, which had been founded in 1877, and founded the Victorian Football League (VFL), which became the sport's major organization, in 1896. There were fourteen clubs (three of which were from states other than Victoria) in the VFL by 1990, when it was superseded by the Australian Football League (AFL). Another club, from Adelaide, joined the AFL in 1991.

FORDE, FRANCIS MICHAEL (1890–1983), politician, was born in Queensland. He was educated at a local state school, and at the Christian Brothers' College at Toowoomba, and worked variously as a teacher, a telegraphist, and an electrical engineer. Forde was a Labor Member of the Queensland Legislative Assembly from 1917 to 1922, and then a Queensland Member of the House of Representatives from 1922 to 1946. He held various portfolios, and was Deputy Prime Minister from 1941 to 1946. After *Curtin's death, Forde was Prime Minister in July 1945 as a short-term measure until *Chifley was elected to that position. Forde served as High Commissioner in Ottawa from 1946 to 1953, after which he returned to state politics, first as a Labor organizer (1953–5), and then as a Member of the Legislative Assembly from 1955 to 1957.

FORREST OF BUNBURY, JOHN, 1ST BARON (1847–1918), politician and explorer, was born in Bunbury, Western Australia, was educated at Bishop Hale's School, Perth, and became a surveyor. He worked in the Surveyor-General's Office from 1865 to 1890, during which time he led several major expeditions, including one to search for *Leichhardt in 1869, and another from Perth to Adelaide in 1870. Forrest was a Member of the Western Australian Legislative Council (1883–90), then a Member of the Legislative Assembly (1890–1901), and was Western Australia's first premier (1890–1901). He was largely responsible for the construction of the Gold-fields Water Supply, which took water to the *Kalgoorlie area, and for the development of the harbour at *Fremantle. Forrest was active in the *federation movement; he attended the National Australasian Convention, several meetings of the Federal Council, and the *Federal Convention. He entered federal politics, was variously a Protectionist, Fusion, Liberal, and Nationalist Member of the House of Representatives from 1901 to 1918, and held a number of portfolios. In 1918 he became the first Australian to be granted a peerage. He died, while in office, at sea off Sierra Leone. His publications include *Explorations in Australia* (1875), *Notes on Western Australia* (1884), and *The Present and Future of Western Australia* (1897).

FORT DUNDAS *Northern Territory; *Port Essington

FORT WELLINGTON *Northern Territory; *Port Essington

'FOURTEEN POWERS' REFERENDUM Popular name given to a referendum in 1944 to alter the *Constitution of the Commonwealth. The subject was post-war reconstruction and democratic rights.

John **Forrest**, seen here in a photograph from the 1890s, was a Western-Australian-born explorer and politician. (Australian Natives' Association)

It was introduced by the *Curtin Labor Government, the proposals having been drafted by *Evatt, during the Second World War. Voters were asked whether they would give the Commonwealth Parliament the exclusive right to legislate in fourteen specific areas for a period of five years after the war ended. State parliaments had previously been responsible for these areas; the Commonwealth Parliament could deal with them during the war. The powers related to the reinstatement of returned service personnel; employment and unemployment; organized marketing of commodities; companies; trusts, combines, and monopolies; profiteering and prices; production and distribution of goods; control of overseas exchange and investment, and raising of money in accordance with Loan Council plans; air transport; uniformity of railway gauges; national works; national health; family allowances; and Aborigines. Voters had to accept or reject all of the powers. The referendum, held on 19 August 1944, was not carried. There were 2 305 418 'No' votes and 1 963 400 'Yes' votes. New South Wales, Victoria, Queensland and Tasmania voted 'No'; South Australia and Western Australia voted 'Yes'.

FOVEAUX, JOSEPH (1765–1846), colonial administrator, pursued a military career. He joined the *New South Wales Corps in 1789, and arrived in New South Wales in 1792. He commanded the Corps during William *Paterson's absence from 1796 to 1799. Foveaux was acting Lieutenant-Governor (1800–1) and then Lieutenant-Governor (1801–4), at *Norfolk Island. He returned to England, partly because of ill health, in 1804. On his return to New South Wales, he found the aftermath of the *Rum Rebellion; *Bligh was under arrest and George Johnston was in control. Foveaux took control from Johnston in mid-1808, and remained in charge until Paterson arrived from Van Diemen's Land in early 1809. Foveaux sailed for England in 1810. His military career continued, and he was promoted lieutenant-general in 1830. He died in London.

FRANKLIN, SIR JOHN (1786–1847), Arctic explorer and colonial administrator, was born and educated in England. His naval career included service in the Battle of Copenhagen (1801), in the Battle of Trafalgar (1805), and off the Greek coast (1830–3). He served on the *Investigator* during *Flinders's *New Holland expedition (1801–3), was second-in-charge of an unsuccessful expedition to find the North-West Passage (1818), and led two expeditions to Arctic America (1819–22, 1824–8). Franklin was Lieutenant-Governor of *Van Diemen's Land from 1837 to 1843, when he was recalled. He then led a naval expedition that discovered the North-West Passage, but he and the rest of the expedition's members died. Relics of the expedition were not found for some years. Franklin's publications include *Narrative of a Journey to the Shores of the Polar Sea, in the Years 1819–22* (1823) and *A Narrative of Some Passages in the History of Van Diemen's Land during the last three years of Sir John Franklin's Administration in the Colony* (1845).

FRANKLIN, STELLA MARIA(N) SARAH MILES (1879–1954), novelist, was born to Australian parents at Talbingo, New South Wales. She was educated privately and at Thornford Public School. While also writing, she worked as a governess near Yea, Victoria, and as a maid in

Sydney and Melbourne. Miles Franklin sailed to the USA in 1906, where she worked in Chicago in secretarial and editorial positions for the National Women's Trade Union League of America, between 1907 and 1915. Alice *Henry was her colleague. In 1915 she sailed to England, where she held various jobs before engaging in voluntary war work with the Scottish Women's Hospitals in Macedonia in 1917 and 1918. She lived in London between 1918 and 1926, and for most of this time held a secretarial position with the National Housing and Town Planning Council. She also visited the USA and Australia in 1923 and 1924. In 1927 she returned to live permanently in Australia, then in 1930–2 visited the USA and England again. During the Second World War she did some voluntary work in Australia. Miles Franklin joined the Fellowship of Australian Writers in 1933 and the Sydney PEN Club in 1935, encouraged other writers, gave lectures on Australian literature, and provided in her will for an annual Miles Franklin award for Australian literature. She died in Sydney. Miles Franklin is best known for her novel *My Brilliant Career* (1901), upon which a later film was based. Her authorship of the pseudonymous 'Brent of Bin Bin' novels was not definitely established until 1966. Her other novels include *Some Everyday Folk and Dawn* (1909), *All That Swagger* (1936), and *My Career Goes Bung* (1946), a sequel to *My Brilliant Career*.

FRASER, JOHN MALCOLM (1930–), politician, was born in Melbourne, was educated at Melbourne Church of England Grammar School and Oxford University, and became a grazier. He was a Victorian Liberal Member of the House of Representatives from 1955 to 1983, and held various portfolios. Malcolm Fraser became Leader of the Liberal Party in 1975, and was appointed caretaker Prime Minister following

*Malcolm **Fraser**, Australia's Prime Minister from 1975 to 1983. (Age)*

the *dismissal of the Whitlam Government in 1975. He was Prime Minister from 1975 to 1983, when his Government was defeated by the Labor Party, led by *Hawke. He then resigned as Leader of the Liberal Party, and from Parliament. His activities since then have included working towards the abolition of apartheid in South Africa.

'FREE, COMPULSORY, AND SECULAR' A popular term used to describe the system of education introduced in the Australian colonies during the latter part of the nineteenth century. All of the colonies passed education Acts to establish national systems of schools which were entirely funded and controlled by the government. These Acts were passed in Victoria in 1872, South Australia and Queensland in 1875, New South Wales in 1880, Tasmania in 1885, and Western Australia in 1893 and 1895. The term is somewhat misleading; some fees were charged, some absences were accepted, and some religion was taught.

FREE TRADE A fiscal policy under which governments theoretically do not intervene directly in the flow of goods and services between, or within, countries. The tariff question was one of the most controversial issues in Australian politics in the late nineteenth and early twentieth centuries. The two most populous colonies, Victoria and New South Wales, adopted different polices—Victoria protectionist and New South Wales free trade. The other colonies adopted various policies. Following federation in 1901, the Constitution of the Commonwealth guaranteed free trade between the states (formerly colonies), but *protection was gradually adopted as a national policy. The *Free Trade Party and the *Protection Party were absorbed into the *Fusion Party in 1909.

FREE TRADE PARTY One of the two main non-Labor political parties (the other being the *Protectionist Party) in the newly formed Commonwealth Parliament in 1901. Many members of the Free Trade Party accepted that some degree of *protection was inevitable, and the Free Trade Party governed federally, in coalition with some Protectionists, in the Reid–McLean Government from 1904 to 1905. In 1906 some of the Free Traders joined with some of the Protectionists and became known as the Tariff Reformers. The Free Trade Party was led by *Reid (1901–8) and Joseph *Cook (1908–9). The Free Traders merged with the Protectionists and the Tariff Reformers to form the *Fusion Party in 1909. There had also been Free Traders in colonial parliaments.

FREMANTLE, a city within the Perth metropolitan area, and Western Australia's main port, is situated on the mouth of the Swan River, where Charles Fremantle took British possession of the western part of Australia in 1829. A settlement was established there later that year. Fremantle was connected by rail with Perth, 19 kilometres away, in 1881; became a municipality in 1883; and a city in 1929. Its inner harbour, designed by Charles O'Connor, was completed in 1897.

FURNEAUX ISLANDS A group of islands in Bass Strait lying north-east of Tasmania, of which they form a part. The larger islands are, in order, Flinders Island (approximately 64 by 29 kilometres), Cape Barren Island, and Clarke Island. The smaller ones include Preservation, Chappell, and Babel Islands. Aborigines are believed to have lived in the area some eight thousand years ago. Tobias Furneaux (after whom James *Cook named the group), commanding one of Cook's ships during the latter's second voyage, sailed along the eastern side of the islands in 1773. The *Sydney Cove* was beached on Preservation Island in 1797, and the following year *Flinders visited the islands twice (once with *Bass). *Sealers began to use these and other Bass Strait islands in the 1790s and early 1800s. Mutton-birding also began on the Furneaux Islands during this period. A settlement for the so-called last Van Diemen's Land Aborigines was found at Wybalenna (or Settlement Point) on Flinders Island in 1832, and continued until 1847, when most of the survivors were moved to Oyster Cove near Hobart. Other settlers gradually arrived on the islands, including *soldier settlers on Flinders Island after the Second World War. Grazing, fishing, and mutton-birding have been the islands' main industries in recent years.

Furphy, Joseph ('Tom Collins') (1843–1912), writer, was born to Irish parents at Yering (present-day Yarra Glen) in the *Port Phillip District. He is more commonly known as 'Tom Collins'. His father was a tenant farmer. Furphy was educated at home and at schools at Kangaroo Ground and Kyneton. His various jobs included operating a threshing plant in the Kyneton district; working a selection in the Colbinabbin district for about six years; owning and driving bullocks in the Riverina district for seven or eight years; and working in his family's iron foundry at Shepparton for about twenty years. He moved to Western Australia in 1904 and died there. Furphy is best known for his novel *Such is Life: Being Certain Extracts from the Life of Tom Collins* (1903). Written in diary form, it is set in the Riverina in the 1880s, and based largely upon his own experiences of bush life. His other publications include *Rigby's Romance* (1921) and *The Buln-Buln and the Brolga* (1948).

Fusion Party The Protectionists led by *Deakin, the Free Traders led by Joseph *Cook, and the Tariff Reformers (or Corner Group) led by *Forrest, merged to become the Fusion Party in the Commonwealth Parliament in 1909. Deakin became the leader, Cook his deputy, and the party governed for a short time in 1909–10. It was known as the *Liberal Party by 1913.

Fuzzy-wuzzy angels A name given by Australians to the Papuans and New Guineans who carried Australian supplies and wounded servicemen, especially along the *Kokoda Trail, during the Second World War.

G

GALLIPOLI CAMPAIGN   An unsuccessful Allied attempt to force the Dardanelles during the First World War. Its main aims were to force Turkey out of the war and to open a safe sea route to Russia. A naval expedition launched in February and March 1915 failed. A military expedition (relying mainly upon British, Australian, New Zealand, and French troops) with some naval support was then attempted. The first landings on the Gallipoli peninsula and the Asian mainland were made on 25 April 1915. The Anzac landing, on a beach later renamed Anzac Cove, is commemorated on *Anzac Day. Turkish resistance was strong. Further landings were made, but fighting on the peninsula reached a stalemate. The Allied troops were withdrawn in December 1915 and January 1916. Australian casualties on Gallipoli were 8587 killed and 19367 wounded.

GAME, SIR PHILIP WOOLCOTT *Lang, John Thomas; *New Guard

GARDINER (CHRISTIE), FRANCIS (1830–1903?), *bushranger, has been described as the 'organizer of outlaws'. He was twice convicted of horse-stealing, before turning to bushranging. Frank Gardiner led the gang which took some £14 000 in gold and notes in the famous gold-escort robbery at Eugowra, *New South Wales, in 1862. He was arrested in *Queensland in 1864. After serving part of his sentence he was released, subject to exile, in 1874. His controversial release led to the downfall of the *Parkes Government. Gardiner left for Hong Kong in 1874. His death in the USA was reported in 1903.

GAWLER, GEORGE (1795–1869), *governor, was born in England, and like his father, pursued a military career. Gawler took office as the Governor of South Australia in 1838; he was also the Resident Commissioner. (The two positions had formerly been separate, and held by *Hindmarsh and James Hurtle Fisher.) Gawler had much land surveyed, undertook extensive public works, and established a police force. At this time South Australia experienced severe financial difficulties, for which Gawler was unfairly blamed. He was recalled in 1841, returned to England, and did not hold public office again. *Grey's dispatches criticized him, but historians have criticized the dispatches. Gawler's publications include *The Close and Crisis of Waterloo* (1813), *The Essentials of Good Skirmishing* (1837), and *The Emancipation of the Jews* (1847).

GEELONG, Victoria's second-largest city and port, is situated at Corio Bay, 74 kilometres south-west of Melbourne by road, at the centre of various rail and road routes. Lieutenant John Murray and Matthew *Flinders both visited the bay in 1802. The following year, William *Buckley began living in the area, which, as members of the *Hume and Hovell expedition noted in 1824, Aborigines called *jillong*, meaning 'white seabird' or 'swamp where native companions live'. Acting on behalf of the *Port Phillip Association (originally the Geelong and Dutigalla Association), John Batman acquired 40 500 hectares of land there in 1835 after signing the Geelong Deed with local Aboriginal leaders. Geelong was declared a town by Governor Richard *Bourke in 1837, the year in which White settlement first began, and was incorporated in 1849. The development of Geelong, which became an important centre for the wealthy Western District, was boosted by the Victorian *gold-rushes of the

GELLIBRAND, SIR JOHN

An early ambrotype showing the laying of the foundation stone of the Geelong clock tower in 1856. (National Gallery of Victoria)

1850s. The town was connected by rail to Williamstown in 1857, Melbourne in 1859, and Ballarat in 1862. Its noteworthy educational institutions include Geelong Church of England Grammar School, established in 1855; Geelong College, established in 1912; and Deakin University, established in 1971. Proclaimed a city in 1912, Geelong continued to develop as a major industrial and commercial centre throughout much of the twentieth century.

GELLIBRAND, SIR JOHN (1872–1945), soldier and farmer, was born at Ouse, Tasmania. His father was a grazier and politician. Gellibrand was educated in England at the King's School, Canterbury, and the Royal Military College, Sandhurst, and pursued a career in the British Army, serving in various places, including South Africa (during the *Boer War) and Ceylon. Soon after retiring, he returned to Tasmania and settled on an orchard at Risdon. Gellibrand served with the first AIF from 1914 to 1919, rising to become the commander of the 3rd Division, with the rank of major-general, in May 1918. He was wounded several times during the First World War, received numerous honours and decorations, including the Distinguished Service Order and Bar, the Croix de Guerre and Legion d'Honneur, and the American Distinguished Service Medal, and is remembered particularly for his contribution to the fighting at Second Bullecourt (May 1917). After the war Gellibrand was the Tasmanian Public Service Commissioner (1919–20), then the Victorian Chief Commissioner of Police (1920–2), and also commanded the 3rd Division of the Australian Military Forces (1921–2). In 1922 he established the Remembrance Club in Hobart, which led to the formation of the Legacy movement (the main aim of which came to be the care of dependants of deceased ex-servicemen). He was a Nationalist Member of the House of Representatives for the Tasmanian seat of Denison from 1925 to 1928. He then farmed in Tasmania for some years, before settling on a property near Yea, Victoria, in 1937.

GIBBS, CECILIA MAY (1877–1969), writer and artist, was born in Sydenham, Kent, England. She, her brothers, and her mother followed her father, a public servant and artist, to Adelaide in 1881. The family moved to Perth, where May attended a girls school. She studied at art and technical schools in England between 1901 and 1904 before returning to Perth, where she contributed articles and illustrations to the *Western Mail*. She again studied and worked as an illustrator in England from 1909 onwards. After returning to Australia in 1913 May Gibbs settled in Sydney's Neutral Bay, where she was to live for the remainder of her life, and continued working as an illustrator. She became famous for the books she wrote and illustrated for children about imaginary bush creatures such as gumnut babies, blossom babies, and the bad banksia men, the first of which was *Gumnut Babies* (1916). *Snugglepot and Cuddlepie* (1918) became immensely popular. She also published various cartoon strips, one of which, 'Bib and Bub', ran in newspapers almost without a break from 1924 until 1967. Her last book was *Prince Dande Lion* (1954). May Gibbs, who had married Bertram James Ossoli Kelly (died 1939) in 1919, but had had no children, died in Sydney. She left her estate to several charities including the United Nations International Children's Emergency Fund. Her other publications include *Boronia Babies* (1917), *Little Ragged Blossom* (1920), and *Mr. and Mrs. Bear and Friends* (1943).

GIBLIN, LYNDHURST FALKINER (1872–1951), political economist, was born in Hobart, the son of William Robert *Giblin. He was educated at the Hutchins School, Hobart; University College, London; and King's College, Cambridge. (He became an honorary fellow of the latter in 1938, and a Giblin studentship was established there after his death.) He also represented England at Rugby Union. Giblin worked at various jobs, including those of gold-prospector and sailor, in different parts of the world, before settling in Tasmania in about 1906. He was a Labor Member of the Tasmanian House of Assembly from 1913 to 1916, then served in the first AIF (having earlier served in the citizen forces). He was wounded several times, was mentioned in dispatches and awarded the Military Cross and Distinguished Service Order, and rose to the rank of major. After the war he was the Tasmanian government statistician from 1919 to 1928, then Ritchie Professor of Economics at the University of Melbourne from 1929 to 1939. During the *Great Depression he was one of the economists who were largely responsible for the so-called Premiers' Plan. He acted as the Commonwealth statistician in 1931 and 1932, was a member of the newly formed *Commonwealth Grants Commission from 1933 to 1936, and of the Commonwealth Bank Board from 1935 to 1942. He chaired an advisory committee on financial and economic policy from 1939 to 1947. Giblin has been described as having 'won professional renown in the fields of Federal finance, tariff policy, employment analysis, Depression diagnosis, and post-war economic management'. He died in Hobart. His publications include *Growth of a Central Bank* (1951), which was published posthumously.

William Dobell's Professor L. F. Giblin *(1945).* **Giblin** *was one of the economists who helped to formulate the Premier's Plan during the Great Depression. (National Library of Australia)*

GIBLIN, WILLIAM ROBERT (1840–87), lawyer and politician, was born in Hobart, and educated at Hobart High School. He became a barrister in 1864, the year in which he also founded the Hobart Working Men's Club. Giblin was a Member of the Tasmanian House of Assembly from 1869 to 1884, was Premier twice (1878, 1879–84), and held other portfolios. He 'reorganized the colony's finances, secured the adoption of an equitable taxation policy and initiated an active programme of public works'. Giblin was appointed to the Tasmanian Supreme Court in 1885. Lyndhurst Falkiner *Giblin was one of his sons.

GILL, SAMUEL THOMAS (1818–80), artist, was born in England and educated at a school run by his father, a Baptist minister, and mother. The family arrived in Adelaide in 1839. Gill established a studio there in 1840, and gained a reputation for his drawings and lithographs. His subjects included rural scenes, views of Adelaide, and local people. In 1846 he was a draughtsman with J. A. Horrock's expedition to the area north of Spencer Gulf, during which Horrocks died. In 1852, during the *gold-rushes, Gill moved to Victoria, and his drawings and lithographs of life on the gold-fields became popular in England and Australia. In 1869 the trustees of the Public Library of Victoria commissioned Gill, who has been called the 'Artist of the Gold-fields', to paint

S. T. Gill's undated water-colour entitled Native Sepulchre *is much more sombre than many of his other works, especially those depicting life during the gold-rushes. (National Gallery of Victoria)*

*William Dobell's Dame Mary Gilmore (1957), painted a few years before **Gilmore**'s death. (Art Gallery of New South Wales)*

forty water-colours of life on the Victorian diggings in the early 1850s. Gill also did some work in New South Wales.

GILMORE (née Cameron), DAME MARY JEAN (1865–1962), writer, was born near Goulburn, New South Wales. Her father, who held farming and other jobs, was Scottish, her mother Australian. She was educated at schools at Brucedale, Wagga Wagga, and Downside, before becoming a pupil-teacher. She taught at schools in New South Wales from 1877 to 1895. She claimed that, although listed under her brother's name, she was the first female member of the *Australian Workers' Union. She was involved in William *Lane's *New Australia movement, and lived at the Cosme settlement in Paraguay from 1896 to 1899, during which time she taught, edited the colony's newspaper, married William Alexander Gilmore in 1897, and bore a son in 1898. Mary Gilmore then taught English in Patagonia, before returning to Australia, via London, with her husband and son. She lived near and in Casterton, Victoria, from 1902 to 1912, when she moved to Sydney, where she remained for the rest of her life. Mary Gilmore wrote for local newspapers while living in the country; her life and poems were featured in the *Bulletin's Red Page on 3 October 1903; she began the *Australian Worker*'s page for women in 1908 and edited it until 1931; she had books of poetry and prose published; she wrote a regular column entitled 'Arrows' for the *Tribune*, a Communist newspaper, from 1952 to 1962; and she was a foundation member of the Fellowship of Australian Writers in 1928, and did much to encourage other writers. She was created a dame in 1937. Her publications include many volumes of poetry, *Hound of the Road* (1922), *Old Days, Old Ways: A Book of Recollections* (1934), and *More Recollections* (1935).

GIPPS, SIR GEORGE (1791–1847), *governor, was born in Kent, England, the son of a clergyman, and educated at the King's School, Canterbury, and the Royal Military Academy, Woolwich. He pursued a military career. He was appointed Governor of New South Wales in 1837, arriving in Sydney the following year. The land proposals of 1844 (the main features being that *squatters would pay an annual licence fee of £10 per run, would be able to buy a 320-acre homestead for £1 per acre after five years' occupation, ensuring tenure of the entire run for eight years, and then would have to buy another 320 acres every eight years to maintain security over the run) were vehemently opposed by the squatters (especially W. C. *Wentworth), whose demands were eventually largely met when the Order-in-Council of 1847 brought the British *Waste Lands Occupation Act* 1846 into operation. Gipps insisted, in the face of much public opposition, upon the prosecution of those involved in the outbreak of violence against Aborigines at *Myall Creek. As instructed, he established the unsuccessful *Port Phillip Protectorate, designed to control and protect Aborigines. He favoured the introduction of a national system of government schools, but ultimately allowed the subsidized denominational system to continue. Gipp's governorship ended in 1846, and he was in very poor health when he left the colony that year.

GLADSTONE COLONY *North Australia

GLOVER, JOHN (1767–1849), landscape painter, was born in England. His father was a farmer. Glover became a writing master at Appleby in about 1789, and a drawing master at Lichfield in 1794. He studied under William Payne, and perhaps John 'Warwick' Smith, in London. In 1805 Glover moved to London, where he taught painting. He was a founding member of the Society of Painters in Water-Colours in 1804, president of it in 1807, and a founding member of the Society of British Artists in 1823. He also painted in oils. Glover lived in the Lake District for two years from about 1817. He opened a gallery for his own works in London's Old Bond Street in 1820. In 1831 Glover settled in *Van Diemen's Land, three of his sons having already done so. He continued to paint landscapes, and in 1835 he exhibited more than sixty pictures, said to be 'descriptive of the Scenery and Customs of Van Diemen's Land', in London. Glover died at his property near Launceston in Van Diemen's Land. His

GOLD-RUSHES

*One of John **Glover**'s best-known landscapes, entitled* The River Nile, Van Dieman's [sic] Land *(1837). (National Gallery of Victoria)*

paintings include *Australian Landscape with Cattle* (c. 1835).

GOLD-RUSHES, rushes of people to newly discovered gold-fields, occurred in several parts of Australia during the nineteenth century. Although gold was discovered earlier, the first rushes did not begin until 1851, when gold was discovered in payable quantities near Bathurst, New South Wales. John Lister and the Tom brothers made the discoveries; Edward Hammond Hargraves publicized them and received official rewards. The richer Victorian goldfields, near Ballarat and Bendigo, were discovered shortly afterwards. Continuing discoveries attracted many people from other countries and other colonies. The population of New South Wales increased from 187 000 to 350 000, and of Victoria from 77 000 to 540 000, between 1850 and 1860. A third of the world's gold output between 1851 and 1861 came from Victoria. Queensland gold discoveries began in the 1860s, and the main rush, to the Palmer River, occurred during the 1870s. Western Australian gold discoveries began in the 1880s; discoveries at Coolgardie in 1892, and *Kalgoorlie in 1893, led to rushes.

GOLDSTEIN, VIDA MARY JANE (1869–1949), feminist and suffragist, was born in Victoria, and educated privately and at the Melbourne Presbyterian Ladies College. She and her sisters ran a preparatory school for part of the 1890s. She became involved in organizations such as the National Anti-Sweating League, helped to organize the Queen Victoria Hospital Appeal for the Queen's Diamond Jubilee in 1897, and campaigned for reformist legislation such as the Victorian *Children's Court Act 1926*. She helped to collect signatures for a female suffrage petition (later presented to the Victorian Parliament) in 1890; spoke publicly in favour of female suffrage from 1899; visited the USA in 1902, attending conferences, giving evidence in support of female suffrage to a Congress committee, and making speeches; helped to found, and

was president of, the Women's Political Association, which existed from 1903 to 1919; founded, owned, and edited the *Woman's Sphere*, a monthly journal; was active in many other women's organizations; founded, owned, and edited the *Woman Voter*, a weekly journal, which existed from 1909 to 1919; went to Britain (invited by the Women's Social and Political Union), where she wrote and spoke advocating female suffrage, and formed the Australian and New Zealand Women's Voters' Association in London in 1911. Goldstein, one of the first women in the British Empire to stand for election to parliament, was unsuccessful as an independent candidate for the Senate in 1903. She stood again for parliament, unsuccessfully, four more times. During the First World War, she chaired the newly formed Peace Alliance; founded, and was president of, the Women's Peace Army, which lasted from 1915 to 1919; and helped to organize a Women's Unemployed Bureau and the Women's Rural Industries Company Limited. She and Cecilia John attended a Women's Peace Conference in Switzerland in 1919. Goldstein returned to Australia about three years later, but did not resume an active public life.

GOLF The first golf-course in the Australian colonies was established in Melbourne in 1847. A golf club was

S. T. Gill recorded many aspects of life during the Victorian gold-rushes, in works such as this lithograph entitled The New Rush *(1863). (National Gallery of Victoria)*

*Vida **Goldstein**, shown here in her 1903 Senate campaign photograph, was one of the first women in the British Empire to stand for election to parliament. (La Trobe Collection, State Library of Victoria)*

founded in Sydney in the 1880s, and others were established there and elsewhere in the 1890s and later. The Australian Golf Union was established in 1898, and the Australian Ladies Golf Union in 1921. Golf has become a popular sport in Australia, especially since the Second World War, and there were more than 1200 golf-courses in the country by the early 1980s. Peter William Thomson (1929–), one of Australia's best-known golfers, won the British Open five times, the Australian Open three times, the New Zealand Open nine times, and numerous other championships.

GORDON, ADAM LINDSAY (1833–70), poet, was born in the Azores. His father was a retired army officer. Gordon was educated in England at Cheltenham College, the Royal Military Academy at Woolwich, and the Royal Worcester Grammar School. He migrated to South Australia in 1853, and belonged to the South Australian Mounted Police from 1853 to 1855. In the remaining years of his life he held various jobs, including those of horse-breaker and steeplechase rider, mostly in South Australia and Victoria. He was a Member of the South Australian House of Assembly in 1865–6. Gordon wrote prose, but is best known for his poetry. He shot himself at Brighton Beach, Victoria, on 24 June 1870, the day after his *Bush Ballads and Galloping Rhymes* was published. Financial difficulties, ill health caused by riding accidents, and the death of his only child apparently were factors leading to his suicide. Gordon's poetry was very popular in the years following his death. His publications include *The Feud: A Ballad* (1864), *Ashtaroth: A Dramatic Lyric* (1867), and *Sea Spray and Smoke Drift* (1867).

GORTON, SIR JOHN GREY (1911–), politician, was born in Melbourne, educated at Sydney Church of England Grammar School, Geelong Grammar School, and Oxford University, and worked for some time on his family's orchard at Mystic Park, Victoria. From 1940 to 1944 he was a fighter pilot in the Royal Australian Air Force and was badly wounded. He was a Victorian Liberal Senator from 1949 to 1968, and then a Liberal Member of the House of Representatives from 1968 to 1975, holding various portfolios. Gorton was Prime Minister from 1968 to 1971, but lost the confidence of his party, using his casting vote to decide the issue, and was replaced by *McMahon. Gorton stood unsuccessfully as an Independent Senate candidate for the Australian Capital Territory in 1975.

GOVERNMENT MEN   A popular name given to convicts during the period of *transportation.

GOVERNOR, JIMMY (1875–1901), outlaw, was born in *New South Wales, the son of a bullock-driver. He was educated at a mission school and at Gulgong, before working in various jobs, including one as a police tracker. While fencing for John Mawbey at Breelong, Governor and another *Aborigine named Jacky Underwood, having allegedly been provoked by racist insults, killed five members of the Mawbey household and wounded another on 20 July 1900. Underwood was captured and hanged, but Jimmy and his brother Joe went bushranging through New South Wales, killed three people and a baby, and

Governors

*John **Gorton**, Australian Prime Minister (1968–71), with Miss Australia in March 1969. (Federal Capital Press of Australia Pty Ltd)*

robbed others. Rewards of £1000 each were offered for their capture. In October Jimmy was captured several days after having been shot, and Joe was shot dead. Jimmy was tried for one of the murders, convicted, and hanged on 18 January 1902. Thomas Keneally's book *The Chant of Jimmy Blacksmith* (1972), and the film of the same name (1978), are based upon Jimmy Governor's experiences.

Governors administered New South Wales from 1788, South Australia from 1836, and Queensland from 1859. At various times lieutenant-governors administered some of the dependencies of New South Wales (which included Van Diemen's Land until 1825). Those appointed to administer some colonies were also originally designated lieutenant-governors; this happened in Western Australia until 1832, and in Van Diemen's Land and Victoria until 1855. Their administrators were then designated governors. Early colonial administrators held very wide powers, but these diminished as other political institutions developed. Two administrators of New South Wales were designated governors-general in the years between 1855 and 1861. Since federation there have been governors-general and state governors. The governors-general of the Commonwealth of Australia have been the Earl of Hopetoun (1901–3), Baron Tennyson (1903–4), Baron Northcote (1904–8), the Earl of Dudley (1908–11), Baron Denman (1911–14), Sir Ronald Craufurd Munro-Ferguson (later Viscount *Novar) (1914–20), Baron Forster (1920–5), Baron Stonehaven (1925–31), Sir Isaac *Isaacs (1931–6), Baron Gowrie (1936–45), the Duke of Gloucester (1945–7), Sir William *McKell (1947–53), Sir William Slim (1953–60), Viscount Dunrossil (1960–1) Viscount De L'Isle (1961–5), Baron *Casey (1965–9), Sir Paul *Hasluck (1969–74), Sir John Kerr (1974–7), Sir Zelman Cowen (1977–82), Sir Ninian Stephen (1982–9), and William George Hayden (1989–). The governor-general represents the Crown in the Commonwealth of Australia, and the governors represent the Crown in the states. Some have deputies, called lieutenant-governors.

GOYDER'S LINE Popular name given to a rainfall line dividing *South Australia. George Woodroofe Goyder, the Surveyor-General, was instructed in 1865 to show 'the line of demarcation between that portion of the country where the rainfall has extended, and that where the drought prevailed'. Goyder's Line stretched from Swan Reach to Mount Remarkable, south to the Broughton River, and north-west to the Gawler Ranges. The area to the north of it, which comprised most of South Australia, was considered to be arid. In theory, Goyder's Line was to be used in the review of pastoral leases; in practice, it was used to show which areas were thought to be suitable for agriculture.

GRAINGER, GEORGE PERCY (from 1911, Percy Aldridge) (1882–1961), pianist and composer, was born in Melbourne, where he learnt piano from his mother, and also studied with Louis Pabst (1892–4). From 1895 to 1899 he studied at the Hoch Conservatory, Frankfurt-on-Main, Germany, where his teachers included Ivan Knorr and James Kwast, and later he studied with Ferruccio Busoni in Berlin. From 1901 to 1914 Grainger lived mostly in London, where his concert career began. He then settled in New York in 1914, served in a United States Army band from 1917 to 1919, became an American citizen in 1918, taught piano at the Chicago Musical College from 1919 to 1928, and chaired the New York University music department in 1932 and 1933. He founded the Grainger Museum at the University of Melbourne in 1935. Grainger toured widely in Europe, Scandinavia, Australasia, and elsewhere. His reputation as a composer was firmly established before the First World War. His compositions, many of which can be described as experimental, include choral, orchestral and chamber works, military band music, and piano pieces (including his best-known work *Country Gardens* (1908)). He composed many settings of British and Danish folksongs, and more than thirty settings of Kipling's poems. Grainger died in New York.

Grants Commission

*The idiosyncratic Australian pianist and composer Percy **Grainger**, photographed in 1934 wearing towelling clothing of his own design. (Grainger Museum, The University of Melbourne)*

Grants Commission *Commonwealth Grants Commission

Great Depression (1930s) A worldwide economic depression, which began in 1929. Australia was badly affected. The main causes of Australia's depression were external—the collapse of wool and wheat prices, and the cessation of overseas loans. Internal factors, including an existing recession and drought, exacerbated the situation. National income fell markedly between 1929 and 1932; unemployment rose to about 30 per cent in 1932, and many other people were only partly employed; wages,

GREAT DEPRESSION

*During the **Great Depression** cartoonist Ted Scorfield predicts pain for poor little Australia, following Uncle Sam's nasty experience, in his cartoon entitled "You're Next!", published in the* Bulletin *on 2 July 1930. (National Library of Australia)*

prices, company profits, dividends, and interest rates fell. The *Australian Labor Party split, the *United Australia Party superseded the *Nationalist Party, and the *Communist Party prospered. Unemployed workers formed many organizations, often associated with trades halls or the Communist Party, and other organizations such as the *New Guard were founded. A *secession movement developed in Western Australia, and new states movements flourished in New South Wales. Some government unemployment relief was made available. There were thousands of evictions, sometimes leading to violence, as on *Bloody Friday; shanty settlements were established; and thousands of people went to the country in a futile search for work. Various possible solutions were suggested, resulting in what is sometimes described as the 'Battle of the Plans'. Sir Otto Niemeyer, a representative of the English banks who visited Australia in 1930 at the Prime Minister's invitation, proposed a deflationary plan involving balanced budgets and cuts in wages and government expenditure. The Labor Federal Treasurer, *Theodore, and the left wing of the Labor Party, proposed a mildly inflationary plan involving the creation of credit. The Labor Premier of New South Wales, J. T. *Lang, proposed an inflationary plan involving repudiation of overseas debts and reduced interest rates. All Australian governments eventually adopted the so-called Premiers' Plan, for which the economists *Copland, L. F. *Giblin, L. G. Melville, and E. O. G. Shann were largely responsible. It was a partly deflationary plan, the aim of which supposedly was 'equality of sacrifice', involving reduced government expenditure, wages, pensions, and interest rates, and increased taxation. It also involved a substantial currency depreciation and some limited borrowing. There has been, and continues to be, debate as to whether it improved or worsened the situation. Recovery began in late 1932. Income, investment, and employment returned to pre-depression levels by 1937. Unemployment was finally nearly eliminated during the Second World War.

GREAT WAR *First World War

GREAT WHITE FLEET Popular name given to the United States Pacific Fleet, which visited Australia in 1908 at *Deakin's invitation during its worldwide tour. The fleet, consisting of sixteen battleships, which were painted white, visited Sydney, Melbourne, and *Albany. It attracted very large crowds, said to have been the largest in Australia's history to that time.

GREENWAY, FRANCIS (1777–1837), architect, was born near Bristol in England. While practising as an architect, he was convicted of forgery and sentenced to death, but the sentence was commuted to fourteen years' *transportation. Greenway arrived in Sydney in 1814, soon received a *ticket of leave, and practised as an architect in the colony. He advised on public works, and *Macquarie appointed him civil architect and assistant engineer in 1816, positions he held until 1822. Greenway received a conditional pardon in 1818 and a free pardon in 1819. He designed numerous notable buildings, some of which still exist, albeit in altered form. His buildings included a lighthouse on the south head of Port Jackson, a *Female Factory at Parramatta, convict barracks in Sydney, St Matthew's Church at Windsor, St Luke's Church at Liverpool, St James's Church in Sydney, and the Supreme Court in Sydney. After his dismissal in 1822 he did little further work, one notable exception being a house in Sydney for *Campbell. Greenway died at his property in the Hunter River Valley.

GREER, GERMAINE (1939–), writer and feminist, was born in Melbourne and educated at the Star of the Sea Convent, Gardenvale, the University of Melbourne (BA Hons, 1959), the University of Sydney (MA Hons, 1962), where she

Souvenir postcard issued to coincide with the visit of the **Great White Fleet** *to Australia in 1908.*

also tutored in English in 1963 and 1964, and Cambridge University (PhD, 1967). She lectured in English at the University of Warwick, England, from 1967 to 1972. Her best-selling study of female stereotypes, *The Female Eunuch* (1969), is considered to have been one of the most influential works to have come out of the international women's liberation movement of the late 1960s and 1970s. Greer, one of Australia's best-known and most controversial expatriates, was a freelance journalist from 1972 to 1979, visiting Professor at the University of Tulsa, Oklahoma, USA, in 1979, and Director of the Tulsa Center for the Study of Women's Literature from 1980 to 1982. Her other publications include *The Obstacle Race: The Fortunes of Women Painters and Their Works* (1979), *Sex and Destiny* (1984), *The Madwoman's Underclothes: Essays and Occasional Writings 1968–85* (1986), *Daddy, We Hardly Knew You* (1989), and *The Change* (1991).

GREY, SIR GEORGE (1812–98), explorer, *governor, and politician, was born in Portugal, and educated in England. He attended the Royal Military College, Sandhurst, joined the British army, and served in Ireland for some years. He sailed to Western Australia in 1837, conducted two expeditions in the north-west, and was the resident magistrate at Albany in 1839–40. He was Governor of South Australia from 1841 to 1845 (following *Gawler). Grey was Governor of New Zealand from 1845 to 1853, Governor of Cape Colony (part of present-day South Africa) from 1854 to 1861, and Governor of New Zealand again from 1861 to 1868. He returned to England, but in 1870 went back to New Zealand where he was a Member of the House of Representatives from 1874 to 1894. He was Prime Minister from 1877 to 1879. Grey was one of New Zealand's representatives at the National Australasian Convention (a significant event in the *federation movement) in 1891. He returned to England in 1894, and died there. His publications include *Vocabulary of the Dialects Spoken by the Aboriginal Races of South-Western Australia* (1839) and *Journal of Two Expeditions of Discovery in North-West and Western Australia during the Years 1837, 1838, and 1839* (1841).

GRIFFIN, WALTER BURLEY (1876–1937), architect and landscape architect, was born near Chicago in the USA. His father was an insurance agent. Griffin studied architecture at the University of Illinois, and was an associate of Frank Lloyd Wright for some years before establishing his own practice. In 1912 he won an international competition for a design for Australia's federal capital, later named *Canberra. His architect wife Marion (née Mahony) had also worked on the plan. From 1913 to 1920 Griffin was the Federal Capital Director of Design and Construction. His term of office was marked by controversy, and various amendments were made, by Griffin and others, to his original plan. In 1920 he refused a place on the Federal Capital Advisory Committee, which took over his role. Griffin also practised privately in Australia, his designs including those for the towns of Griffith and Leeton; housing estates such as Summit and Glenard (both at Eaglemont) in Melbourne, and Castlecrag (the construction of which was never completed) in Sydney; Newman College, at the University of Melbourne,

Phil May's Dishing his Enemies (1888). *Sir Samuel Griffith was Premier of Queensland twice, 1883–6 and 1890–3.* (La Trobe Collection, State Library of Victoria)

Melbourne's Capitol Theatre, and numerous municipal incinerator buildings. In 1935 Griffin left Australia to work in India, where he later died.

GRIFFITH, SIR SAMUEL WALKER (1845–1920), politician and lawyer, was born in Wales, educated at schools in Ipswich, Woolloomooloo, and Maitland, and at the University of Sydney. He was admitted to the Bar in Brisbane in 1867, and was appointed Queen's Counsel in 1876 and a Privy Counsellor in 1901. Griffith was a Member of the Legislative Assembly of Queensland from 1872 to 1893. He was Premier twice (1883–6,

1890–3) and held numerous other portfolios. His first government passed legislation aimed at allowing small settlers, rather than *squatters, greater access to the land. It passed legislation banning the importation of Pacific Islanders (brought to Australia under the *blackbirding practice) after the end of 1890, but Griffith lifted the ban in 1892. During the *Shearers Strike of 1891, Griffith, who previously had received some support from the labour movement, used soldiers to defeat the strikers, some of whom were imprisoned. A strong advocate of federation, Griffith drafted the bill for the constitution of the Federal Council of Australasia; presided over the Federal Council three times; attended the Australasian Federation Conference in 1890; and was largely responsible for the draft constitution (much of which later became part of the *Constitution of the Commonwealth of Australia) produced by the National Australasian Convention of 1891. He was Chief Justice of Queensland from 1893 to 1903, and the first Chief Justice of the *High Court of Australia from 1903 to 1919. Griffith strongly defended states' rights while on the High Court.

GROSE, FRANCIS (1758?–1814), soldier and colonial administrator, was born in England. He pursued a military career, and helped to raise, and commanded, the *New South Wales Corps. He was appointed Lieutenant-Governor of New South Wales in 1789, and arrived there in 1792. When ill health forced *Phillip to leave the colony later in that year, Grose took over. He controlled the colony for much of the interregnum. In 1794 ill health forced him, too, to leave; William *Paterson then took control. Grose's military career continued, and he died in England.

GUÉRARD, JOHANN JOSEPH EUGEN VON (1812?–1901), artist, was born in Vienna, and studied art in Italy and Germany for a number of years. His father was a court painter. After leaving Germany in 1848, he may have visited the Californian gold-fields. He arrived in Victoria in late 1852, and spent about a year on the gold-fields, where he made numerous sketches. In the following years he travelled and sketched extensively (sometimes with scientific expeditions, such as A. W. Howitt's of 1860) in Australia and New Zealand. He was the curator of the Melbourne National Gallery, and the first master of its art school, from 1870 to 1881, when he resigned for health reasons. His landscape paintings, said to have been influenced by the Austrian Biedermeier School, were exhibited widely in Australia, Europe, and elsewhere. He illustrated S. D. Bird's *On Australasian Climates* (1863), and published a collection of lithographs of Western District scenes entitled *Australian Landscapes* (1867). His paintings include *Valley of the Mitta Mitta, with the Bogong Ranges* (1866).

H

HALL, BENJAMIN (1837–65), *bushranger, was the son of ex-convicts. Although questioned over his part in *Gardiner's famous gold-escort robbery at Eugowra, he was not committed for trial. Ben Hall, who has been described as the most efficient of the bushranger leaders, led various gangs, including one which operated on the Sydney–Melbourne road in 1864. In one notable incident, the gang robbed sixty travellers at Jugiong. Hall's exploits led to the introduction of the New South Wales *Felons' Apprehension Act* 1865, which outlawed bushrangers. He was betrayed, ambushed, and killed by police in 1865.

HAMPTON, JOHN STEPHEN (1810?–69), *governor, served as a British naval surgeon before becoming Comptroller-General of convicts in *Van Diemen's Land in 1846, a position that he held until he left the colony in 1855. He was involved in controversy when he would not give evidence to a Legislative Council Select Committee inquiring into convict administration in 1855. During Hampton's term of office as Governor of Western Australia, which lasted from 1862 to 1868, he treated convicts strictly; he used convict labour extensively for public works, such as the building of Government House in Perth, the Perth Town Hall, and numerous roads and bridges, including a bridge over the river at Fremantle; and he helped to improve the colony's financial position. He died in England.

HANCOCK, SIR WILLIAM KEITH (1898–1988), historian, was born in Melbourne, and educated at the Universities of Melbourne and Oxford. He was Professor of History at Adelaide University (1924–33), Professor of History at Birmingham University (1934–44), Professor of Economic History at Oxford (1944–9), Director of the Institute of Commonwealth Studies at London University (1949–57), and Professor of History and Director of the Research School of Social Sciences at the Australian National University (1957–61). He edited the British official histories (civil series) for the Second World War, and wrote *British War Economy* (1949) with M. M. Gowing. Sir Keith Hancock's other publications include *Australia* (1930), *Smuts: The Sanguine Years 1870–1919* (1962), *Smuts: The Fields of Force 1919–1950* (1968), and *Discovering Monaro* (1972).

HARE-CLARK SYSTEM An unusual form of proportional representation based upon the single transferable vote, designed to provide for the representation of minorities, used in elections for the Tasmanian House of Assembly. The state is divided into a number of electorates, each of which returns multiple members. A seat is allocated when a quota of votes is reached. The quota is determined by dividing the number of valid ballot papers by the number of candidates to be elected plus one, then adding one. After three unsuccessful attempts, Andrew Inglis *Clark, influenced by the ideas of the English barrister and electoral theorist Thomas Hare, introduced the system for a trial period in Hobart and Launceston in 1896. It was suspended in 1902, but was reintroduced for all of Tasmania in 1907. The system did not provide for separate by-elections until a change was made in the 1980s.

HARGRAVE, LAWRENCE (1850–1915), aeronautical pioneer, was born and educated in England. After migrating to Sydney in 1865 he worked in engineering jobs, took part in several expeditions to New Guinea and elsewhere, and was an assistant astronomical observer at the Sydney Observatory for some years. From 1883 onwards he worked full-time on aeronautical and other research, which included construction of model aeroplanes, experimentation with box-kites (with which he lifted himself from the ground in 1894), and development of aeronautical engines. His research had some influence in other countries.

HARGRAVES, EDWARD HAMMOND *Gold-rushes

HARNEY, WILLIAM EDWARD (1895–1962), writer, received little formal education. He worked in the bush until his enlistment in the *Australian Imperial Force in 1915. After returning to Australia after the First World War, Harney worked as a drover and cattleman. He came into close contact with Aborigines in remote parts of northern Australia, and was a native-affairs officer for some time; later, he was curator of Ayers Rock (now Uluru). Bill Harney's radio talks and books about bush life became well known. His books include *Taboo* (1943), *Brimming Billabongs* (1947) and (with A. P. Elkin) *Songs of the Songman* (1949).

HARPUR, CHARLES (1813–68), poet, was born at Windsor, New South Wales, where his father was a schoolmaster and parish clerk. Both of Harpur's parents had been transported to Australia as convicts. Harpur held various jobs, including those of clerk, farmer, and assistant gold commissioner. Harpur is best known for his poetry. His publications include *Songs of Australia* (1851), *The Bushrangers: A Play in Five Acts* (1853), and *The Tower of the Dream* (1865). He also wrote literary and political criticism.

HARRIS, ALEXANDER (1805–74), writer, was born in London. His father was a Nonconformist minister and schoolmaster. Harris worked as a compositor in London, before coming to Australia in 1825, where he held various jobs. After some fifteen years he returned to England where he was a missioner in London for some time. In 1851 Harris migrated to the USA where he taught, wrote, and preached. He died in Canada. Harris is best known in Australia for *Settlers and Convicts: or Recollections of Sixteen Years' Labour in the Australian Backwoods, by an Emigrant Mechanic* (1847), which is a partly factual, partly fictitious account of life in New South Wales during the 1820s and 1830s. His other publications include *The Emigrant Family: or The Story of an Australian Settler* (1849).

HARRIS, MAX *Angry Penguins; *Ern Malley hoax

HARTOG (or Hartooch or Hartichs), DIRCK (or Dirk) (dates unknown), Dutch mariner, is the first European known to have landed on the western coast of Australia. He did so in October 1616, while commanding the *Eendracht*, a Dutch East India Company ship, on a voyage from Holland to Batavia (present-day Jakarta). He examined part of the coast, landed on an island (now named Dirk Hartog Island) near Shark Bay, and left behind an inscribed pewter plate. The area nearby became known as Eendracht Land. William de Vlamingh, another Dutch mariner, replaced the plate with another (now known as the Vlamingh Plate) in 1697.

*The plate left by Dirck **Hartog** on an island off the Western Australian coast in 1616 was replaced by the Vlamingh Plate, shown here, in 1697. (National Library of Australia)*

HARVESTER JUDGEMENT Popular name given to the judgement of 1907 that introduced the principle of the *basic wage for men (also known as the family, living, or minimum wage). A Victorian manufacturer, Hugh Victor *McKay, applied for exemption from excise duties on agricultural machinery made by his Harvesting Machine Company, on the grounds that he paid his employees 'fair and reasonable' wages. Such exemption was possible under the *Excise Tariff (Agricultural Machinery) Act* 1906, which was part of the *New Protection legislation. The President of the *Commonwealth Conciliation and Arbitration Court, Henry Bournes *Higgins, ruled that McKay was not paying 'fair and reasonable' wages. Higgins calculated that a male employee maintaining a household of 'about five persons' (that is, two adults and three children) required wages of seven shillings per day, or forty-two shillings per week, to cover such costs as food, clothing and shelter. McKay was paying unskilled male labourers six shillings per day.

HASLUCK, SIR PAUL MEERNAA CAEDWALLA (1905–), governor-general, was born in Fremantle, Western Australia, and educated at the Perth Modern School and the University of Western Australia. He worked as a journalist with the *West Australian* newspaper from 1922 to 1938, and as a lecturer (later, reader) in the History Department of the University of Western Australia (1939–40, 1948–9). He worked in the Commonwealth Department of External Affairs from 1941 to 1947. Hasluck also entered federal politics; he was a Liberal Member of the House of Representatives from 1949 to 1969, serving as Minister for Territories (1951–63), Minister for Defence (1963–4), and Minister for External Affairs (1964–9). His term of office as Minister for External Affairs, during which he encouraged local participation in the administration of *Papua New Guinea, was particularly noteworthy. He was Governor-General of the Commonwealth of Australia from 1969 to 1974. Hasluck's publications include *Black Australians* (1942), *Workshop of Security* (1948), *Native Welfare in Australia* (1953), *The Government and the People, 1939–45* (two volumes, 1951, 1970), part of the Australian official war history, and *The Office of Governor-General* (1979).

HAWKE, ROBERT JAMES LEE (1929–), politician, was born in Bordertown, South Australia, the son of a Congregational minister. He attended Perth Modern School, graduated BA and LLB from the University of Western Australia, and was Western Australia's Rhodes Scholar of 1952. Hawke was a research officer and industrial advocate for the *Australian Council of Trade Unions from 1958 to 1969, president of that organization from 1970 to 1980, and national president of the *Australian Labor Party from 1973 to 1978. He became a Labor Member of the House of Representatives in 1980. During his period as Prime Minister, which lasted from 1983 to 1991, the ALP made a controversial shift to the middle ground of politics. Although he had been the most electorally successful federal Labor leader to that time, Hawke was replaced by *Keating in 1991.

HAYNES, JOHN *Archibald, Jules François; *Bulletin*

HEAGNEY, MURIEL AGNES (1885–1974), trade unionist and feminist, was born in Brisbane and educated at a convent in Richmond, Melbourne. Her father, a publican and later a carpenter, was active in the labour movement. She worked as a primary teacher until 1915 when she became a clerk in the Defence Department, receiving *equal pay because she was the only woman. She was a member of the Political Labor Council's Richmond branch from 1906, was a delegate to the Women's Central Organizing Committee in 1909, attended the first Victorian Labor Women's Conference, joined the committee of the Workers' Education Association in 1915, and campaigned against *conscription. After the *First World War she worked on the submission of the Federal Unions of Australia to the Commonwealth Cost of Living (Basic Wage) Royal Commission in 1919 and 1920, was secretary of the Australian Relief Fund for Stricken Europe from 1921 to 1923, and then spent two years abroad, visiting Russia, Geneva, where she worked for the International Labour Organization (ILO), and London, where she attended the first British Commonwealth Labour Conference. Muriel Heagney was a member of the Victorian central executive of the *Australian Labor Party in 1926 and 1927, and attended the first Pan-Pacific Women's Conference in Honolulu, presenting a paper on trade-union women. During the *Great Depression of the 1930s she founded the Unemployed Girls' Relief Movement in 1930 and published *Are Women Taking Men's Jobs?* in 1935. From 1936 to 1942 Heagney worked as a travel organizer in Sydney for the Queensland Tourist Bureau. In 1937 she helped to found the Council of Action for Equal Pay, under the auspices of the Federated Clerks' Union, and was its honorary secretary from 1937 to 1949. In 1941 she attended the ILO Conference in New York. She worked as an organizer for the Amalgamated Engineering Union in Sydney from 1943 to 1947. After returning to Melbourne, she became the secretary of the Women's Central Organizing Committee, and a member of the Labor Party's central executive in 1955. Her other publications include *Equal Pay for the Sexes* (1948) and *Arbitration at the Cross Roads* (1954).

HEIDELBERG SCHOOL A term that refers to a group of Australian impressionist painters, since described by Bernard Smith as 'the first distinctive Australian school of painting', that flourished during the late 1880s. The group's name came from Heidelberg, the area of Melbourne that inspired many of their paintings, and the leading painters were *Conder, *McCubbin, *Roberts, and *Streeton. The most famous exhibition of Heidelberg School paintings was the *Exhibition of 9 x 5*

*Tom Roberts, considered to be the founder of the **Heidelberg School**, painted* The Artists' Camp *(c. 1886) while he and Frederick McCubbin were camped near Housten's farm at Box Hill. (National Gallery of Victoria)*

Impressions (painted on cigar-box lids measuring 9 by 5 inches), which was held in Melbourne in 1889. Examples of Heidelberg School paintings include Conder's *The Departure of the S. S. Orient—Circular Quay* (1888) and Streeton's *Still Glides the Stream and Shall Forever Glide* (1889). Many were of distinctive, sometimes historical, Australian subjects; examples include McCubbin's *Down on His Luck* (1889) and Roberts's *Shearing the Rams* (1889–90).

HEINZE, SIR BERNARD THOMAS (1894–1982), music scholar and conductor, was born at Shepparton, Victoria, and studied at the Conservatorium attached to the University of Melbourne, London's Royal College of Music (of which he was later made a Fellow), and the Schola Cantorum, Paris. He served in the Royal Artillery from 1915 to 1920. Heinze was the Ormond Professor of Music at the University of Melbourne from 1925 to 1956, then director of the New South Wales Conservatorium, Sydney, from 1956 to 1966. He became conductor of the Melbourne University Symphony Orchestra in the mid-1920s, was appointed life conductor of the Melbourne Philharmonic Society in 1928, was conductor of the Victorian Symphony Orchestra from 1933 to 1956, and also appeared as guest conductor in various other countries. In 1925 he inaugurated 'Young People's Concerts' (a forerunner of the Australian Broadcasting Commission Youth Concerts, introduced in 1947, for which he was also largely responsible). Heinze, who became musical adviser to the Australian Broadcasting Company in 1929, and to its successor, the *Australian Broadcasting Commission, in 1934, did much to develop the presentation of orchestral concerts in Australia.

HENRY, ALICE (1857–1943), journalist and feminist, was born at Richmond, Melbourne. Her parents were Scottish, her father an accountant, her mother a seamstress. Alice Henry attended various Melbourne schools, including Richard Hale Budd's Educational Institute for Young Ladies. She taught for some time, before becoming a journalist in 1884. For some twenty years she wrote for journals and newspapers such as the *Argus* and the *Australasian*. She campaigned for social reforms such as the introduction of female *suffrage in Australia (and later in America). After visiting England in 1905, Alice Henry settled in the USA. She lectured, wrote, and organized for the National Women's Trade Union League of America in Chicago; she edited the women's page of the *Union Labor Advocate* (1908–19), and the League's journal, *Life and Labor* (1911–15); she ran the League's Educational Department (1920–2); and under the League's auspices, she visited England, Europe, and Australia (1924–6). After returning to live in Melbourne in 1933 she was active in such organizations as the *National Council of Women. Alice Henry died at Malvern, Melbourne. Her publications include *The Trade Union Woman* (1915) and *Women and the Labor Movement* (1923). In 1937 she compiled a bibliography of Australian women writers.

HENTY, EDWARD (1810–78), pastoralist and politician, was born in Sussex, England. His father was a farmer and banker. In 1832 he went to Launceston, in Van Diemen's Land, with his parents and some other members of the family. In November 1834, accompanied by

Henry Camfield and four indentured servants, he began to establish a settlement at Portland Bay (on the coast of present-day Victoria), which both he and his father had previously visited. Edward, who has generally been considered to be the first permanent White settler in the *Port Phillip District, was soon joined by several of his brothers. The Hentys farmed, were involved in whaling, and also took up runs inland. The British Government refused their requests, which began before Edward settled at Portland Bay, to buy land on the mainland. After the proposed town of Portland was surveyed in 1839, the Hentys had to buy land at auction and to remove some of their improvements. They later reached a settlement with the authorities. Edward Henty moved to Muntham, one of the family's stations on the Wannon River, in the early 1840s, and was a Member of the Victorian Legislative Assembly from 1856 to 1861. He died in Melbourne.

HERBERT, SIR ROBERT GEORGE WYNDHAM (1831–1905), politician, was born in Brighton, England, and educated at Eton and Balliol College, Oxford. His father was a barrister. Herbert was private secretary to W. E. Gladstone (then Chancellor of the Exchequer) in 1855, and was called to the Bar of the Inner Temple in 1858. He arrived in the newly separated colony of Queensland as private secretary to Governor Bowen and as Queensland's first colonial secretary in 1859. Herbert was Queensland's first premier, from 1860 to February 1866 (and again briefly in July and August 1866), during which time his main aims were 'to extend settlement especially on the north coast, encourage immigration, diversify the economy and establish a firm basis for government'. Herbert returned to England in 1866 and became Assistant Secretary to the Board of Trade (1868–70), Assistant Under-Secretary (1870–1), and Permanent Under-Secretary (1871–92, and for a brief period some years later) in the Colonial Office. He was Agent-General for Tasmania from 1893 to 1896, and chaired the Tariff Reform Commission in 1903. He died in England.

HIGGINS, HENRY BOURNES (1851–1929), lawyer and politician, was born and educated in Ireland, and arrived in Melbourne with his family in 1870. His father was a Methodist minister. Higgins taught in schools and privately, and studied arts and law at the University of Melbourne (on the council of which he later served from 1887 to 1923). He became a leading equity lawyer, and was appointed King's Counsel in 1903. He was a Member of the Victorian Legislative Assembly from 1894 to 1900, when he was defeated, partly because of his opposition to Australia's involvement in the *Boer War. Higgins represented Victoria at the *Federal Convention (1897–8), and opposed the subsequent Bill. He was a Member of the House of Representatives from 1901 to 1906, and served as Attorney-General in the short-lived *Watson Government of 1904, although he was not a member of the Labor Party. Higgins was appointed to the *High Court in 1906, and became the President of the *Commonwealth Conciliation and Arbitration Court the following year. His best-known decision was the *Harvester Judgement. He resigned as President in 1920, but remained on the High Court until his death. He was a strong advocate of Irish Home Rule. His publications

*Henry Bournes **Higgins**, photographed at the time of the Federal Convention of 1897–8, at which he represented Victoria. (Australian Natives' Association)*

include *Essays and Addresses on the Australian Commonwealth Bill* (1900) and *A New Province for Law and Order* (1922).

HIGH COURT Australia's federal supreme court. It interprets the Constitution, has an original jurisdiction, for example, for matters arising under treaties, and is the final court of appeal from state and federal courts. It was created by Section 71 of the *Constitution of the Commonwealth, and its first members were appointed in 1903; they were the Chief Justice, Sir Samuel *Griffith, and the two puisne judges, Sir Edmund *Barton and Richard Edward O'Connor. The size of the Court was increased to five in 1907, and to its present size of seven in 1912. The Chief Justices of the High Court have been Sir Samuel Griffith (1903–19), Sir Adrian Knox (1919–30), Sir Isaac *Isaacs (1930–1), Sir Frank Gavan Duffy (1931–5), Sir John *Latham (1935–52), Sir Owen *Dixon (1952–64), Sir Garfield *Barwick (1964–81), Sir Harry Gibbs (1981–7), and Sir Anthony Frank Mason (1987–). The High Court has been based in Canberra since 1980.

HIGINBOTHAM, GEORGE (1826–92), politician and lawyer, was born in Dublin. After migrating to Australia in 1854 he pursued careers in the law, journalism, and politics. Higinbotham edited the Melbourne *Argus* from 1856 to 1859. He was then a Member of the Victorian Legislative Assembly (1861–71, 1873–6), and from 1863 to 1868 he was Attorney-General. He is considered to have been Victoria's leading radical at that time. In 1866 he chaired a royal commission inquiring into public education. Higinbotham was appointed to the Victorian Supreme Court in 1880, becoming Chief Justice in 1886. Parliament publicly thanked him for his work in consolidating the Victorian statutes.

HINDMARSH, SIR JOHN (1785–1860), *governor, pursued a career in the British navy before becoming the first governor of South Australia in 1836. He disagreed with *Light's choice of site for the capital, Adelaide. He and James Hurtle Fisher, the Resident Commissioner (who represented the Colonization Commission), also clashed over various issues, and Hindmarsh was recalled in 1838. *Gawler took office as both Governor and Resident Commissioner later that year. In 1840 Hindmarsh was appointed Governor of Heligoland, a position he held until 1856. He died in London.

HINKLER, HERBERT JOHN LOUIS (1892–1933), aviator, was born and educated in Bundaberg, Queensland. His father was a German stockman. While young, Hinkler studied mechanics, built gliders, and worked as a mechanic for an American aviator in Australia and New Zealand. He also worked in the Sopwith factory in England for some time. During the First World War he enlisted with the British Royal Naval Air Service, served variously as mechanic, gunner, and pilot, and was awarded a Distinguished Service Medal. He then worked for A. V. Roe & Co. in England, bought an Avro Baby from them, and made a record non-stop flight from England to Italy in it in 1920. Unable to continue to Australia as planned, he returned to England. He made a record non-stop flight in the Avro Baby (which had been shipped to Australia) from Sydney to Bundaberg in 1921. After returning to England, Hinkler was chief test pilot at A. V. Roe & Co. until 1926. He made a non-stop flight in his Avro Avian from

England to Latvia in 1927, then made the first solo flight from England to Australia in the Avro Avian in 1928, the total time for the trip being fifteen days and two and a quarter hours. Hinkler was unable to find backing to produce the Ibis amphibian aircraft, which he designed and built. He flew solo in his Puss Moth from Canada to New York and then to London (via several countries) in 1931, creating more records. Hinkler crashed and died in the Italian Alps in 1933, while attempting to fly solo in his Puss Moth from England to Australia.

HOBART (officially Hobart Town until 1881), the capital city of the state of *Tasmania, is situated on the estuary of the Derwent River. It was founded as a convict settlement in 1804 and named after the British Secretary of State for the Colonies. It was proclaimed a city in 1842, and incorporated in 1857. Its population was 179 900 in 1988.

HOLMAN, WILLIAM ARTHUR (1871–1934), politician, was born in London. He, his parents (who were both actors), and his brother came to Australia in 1888. A cabinet-maker by trade, Holman became active in the New South Wales labour movement. He became a Member of the New South Wales Legislative Assembly in 1898, and was a vocal opponent of the *Boer War. Holman represented the Labor Party until 1916, when he was expelled from the party because of his support for *conscription, and the Nationalists from then until 1920, when he lost his seat. He was Premier of New South Wales from 1913 until 1920, and held various other portfolios. Holman had been admitted to

A. C. Cooke's Hobart Town *(1879), showing Mt Wellington in the background.* **Hobart** *was officially known as Hobart Town until 1881. (National Library of Australia)*

*Bernhardt **Holtermann**'s Panorama of Sydney Harbour and Suburbs (1875). (National Gallery of Victoria)*

the Bar in 1903, was appointed King's Counsel in 1920, and became a prominent barrister in the 1920s. He also entered federal politics, and was a Member of the House of Representatives for the *United Australia Party from 1931 until his death.

HOLT, HAROLD EDWARD (1908–67), politician, was born in Sydney, educated at Wesley College, Melbourne, and the University of Melbourne, and practised as a solicitor briefly from 1933. He was a Victorian Member of the House of Representatives from 1935 until 1967, representing the *United Australia Party and its successor, the *Liberal Party of Australia. He held various portfolios at different times. During the Second World War Holt enlisted with the second AIF, but was recalled to the ministry. He was Deputy Leader of the Liberal Party from 1956 to 1966. When *Menzies retired in 1966, Holt became Prime Minister, at a time when Australia's involvement in the *Vietnam War was growing. While in office, Holt disappeared while swimming near Portsea, Victoria, on 17 December 1967. Although his body was never recovered, he was presumed dead.

HOLTERMANN, BERNHARDT OTTO (1838–85), gold-miner, merchant, and sponsor of photography, was born and educated in Hamburg, Germany. He worked in his family's mercantile firm before migrating to Australia in 1858. The following year he and Louis Beyers went to the Hill End gold-fields in New South Wales. Unsuccessful for some years, during which Holtermann also worked as a hotel licensee and at various other occupations, they eventually found rich veins at Hawkins Hill in 1871. Their Star of Hope Gold Company was floated in April 1871 and on the night of 19–20 October 1872 the world's largest specimen of reef gold was found on their claim. 'Holtermann's Nugget', as it became known, weighed about 286 kilograms. Holtermann built a house at St Leonards in Sydney and sponsored a collection of photographs of the towns and gold-fields of New South Wales and Victoria by Beaufoy Merlin and Charles Bayliss. The famous Holtermann panorama (1875), providing a continuous view of Sydney Harbour and the surrounding suburbs, was taken from the tower of his house. Some of the photographs were exhibited, in an effort to encourage *immigration, at international exhibitions at Philadelphia and Paris in 1878, winning a number of awards. Holtermann also exhibited some of them elsewhere in the USA and Europe, before returning to Australia with various agencies for German products, including 'Life Preserving Drops'.

He was a Member of the New South Wales Legislative Assembly from 1882 until his death at St Leonards, his main interests being immigration and the idea of a north shore bridge.

HORDERN, ANTHONY (1819–76), merchant, was born in London, and brought to Sydney by his family in 1826. He was educated by J. D. *Lang, and then lived in Melbourne for several years, before he and his brother Lebbeus (1826–81) established a draper's store on Brickfield Hill, Sydney, in the mid-1840s. Anthony established his own retail business in the Haymarket in 1855, his firm becoming the now famous Anthony Hordern & Sons after his sons Anthony (1842–86) and Samuel (1849–1909) became partners in 1865 and 1869. After his death, they continued to develop the business. In 1912 Anthony Hordern & Sons became a private company, with Samuel's son, Sir Samuel (1867–1956) as governing director, and in 1926 it became a public company.

HORSE-RACING, in some form, began in Australia soon after White settlement. Recorded race meetings were first held in New South Wales in 1810, Van Diemen's Land in 1814, Western Australia in 1833, South Australia and the *Port Phillip District in 1838, and the *Moreton Bay District in 1843, and horse-racing has since become a major sport and industry in Australia. The *Melbourne Cup has been the country's best-known horse-race for many years. In 1930 it was won by the now-legendary *Phar Lap, ridden by James Pike (1892–1969). Other notable Australian jockeys have included David Hugh ('Darby') Munro (1913–66), Arthur ('Scobie') Breasley (1914–) and George Moore (1923–).

HOTHAM, SIR CHARLES (1806–55), *governor, was born in England. His father was an honorary canon. Hotham pursued a naval career, and negotiated a British commercial treaty with Paraguay in 1852. He administered Victoria from June 1854 until his death in December 1855, first as Lieutenant-Governor, and then as Governor from May 1855. He sent his resignation to England in November 1855. Hotham's administration of the colony, particularly his handling of the *Eureka Rebellion, has been criticized.

HOWE, MICHAEL *Bushrangers

HUGHES, WILLIAM MORRIS (1862–1952), politician, was born in London, grew up in Wales and England, and became a schoolteacher. He migrated to Australia in 1884, where he held various jobs. Hughes was a Labor Member of the

HUGHES, WILLIAM MORRIS

David Low's cartoon character 'Billiwog', based upon Billy Hughes, from The Billy Book, Hughes Abroad, Cartoons by Low *(1918).*

Legislative Assembly of New South Wales from 1894 to 1901, belonged to the Socialist League from 1892 to 1898, was active in union organization of waterside workers, and was a union official from 1899 to 1916. He was admitted to the Bar in 1903. Hughes was a Member of the House of Representatives from 1901 to 1952, representing Labor from 1901 to 1916, National Labor in 1916–17, and then the Nationalists, the *United Australia Party, and the *Liberals. (He was expelled from the Labor Party over *conscription, from the Nationalists when he helped to bring down the government in 1929, and from the United Australia Party when he rejoined the Advisory War Council, which the party had left, in 1944.) Hughes was Prime Minister from 1915 until 1923, when *Bruce replaced him as leader of the Nationalists and Prime Minister. During the First World War he tried twice (unsuccessfully both times) to introduce conscription. He attended meetings of the Imperial War Cabinet and Imperial War Conference in 1918, and achieved some success when he sought separate representation for the dominions at the *Paris Peace Conference, held after the war. At the latter, he argued effectively for war reparations for Britain and Australia, succeeded in gaining a mandate over German New Guinea for Australia, and defended the *White Australia Policy, opposing a Japanese attempt (which ultimately failed) to include a guarantee of the equality of nations in the covenant of the *League of Nations. Hughes held various other portfolios, and led the United Australia Party from 1941 to 1943. He wrote a number of books.

HULKS The name given to bodies of dismantled ships, worn out and unfit for sea service, which were used as prisons. Convicts had been transported to the British colonies in America until the American War of Independence. Hulks on the River Thames (and later elsewhere) were then used to house convicts, supposedly as a temporary expedient until transportation could be resumed. They rapidly became overcrowded. The need for a place to which convicts could be transported was a major factor in the British Government's decision to establish a settlement at *Botany Bay. Britain continued to use hulks until the mid-nineteenth century. They were also used in Sydney, Hobart and Melbourne.

HUME AND HOVELL EXPEDITION (1824–5) An important feat of inland exploration. In October 1824 Hamilton Hume (1797–1873) and William Hilton Hovell (1786–1875) left the former's station at Gunning, near present-day Yass, in New South Wales. Their destination was Western Port, in present-day Victoria. During their journey they discovered numerous rivers, including the Hume River (later renamed the Murray), and lands suitable for grazing and agriculture. In December 1824 they arrived at Corio Bay in Port Phillip, which they mistook for Western Port, because of a miscalculation made by Hovell. They followed a similar route back to Gunning, arriving there in January 1825. In 1826 Governor *Brisbane sent a party, including Hovell, to Western Port by sea to establish a settlement. Hovell discovered his mistake, and the settlement at Western Port, which was found to be unsuitable, was abandoned shortly afterwards. Hume and Hovell later disagreed over who deserved credit for leading the expedition, and whether or not Hume had mistaken Port Phillip for Western Port at the time.

HUNTER, JOHN (1737–1821), *governor, was born at Leith in Scotland. He considered a clerical career in the Church of Scotland, and briefly attended the University of Aberdeen as an undergraduate, but rejected it in favour of a naval career, during which he had extensive sea-going experience. He sailed to New South Wales with the *First Fleet as second captain, under *Phillip, of HMS *Sirius*, and commanded the main convoy after Phillip transferred to the tender *Supply* during the voyage. Hunter

William Minehead Bennett's Vice Admiral John Hunter, Governor of New South Wales, 1795–1800 *(1813)*. **Hunter** *was the new colony's second governor. (Dixson Galleries, State Library of New South Wales)*

OPPOSITE PAGE
Frank **Hurley**'s *photograph* We took a dog team down to the wreck and salvaged a few essentials *(1915) captures some of the drama of the* Endurance *breaking up. (Australian National Gallery)*

held a dormant commission to succeed Phillip in case of death or absence, but was not Lieutenant-Governor. He was involved in survey and legal work in the new colony, and also took the *Sirius* to the Cape of Good Hope for supplies (1788–9). After the *Sirius*, under Hunter's command, was wrecked off *Norfolk Island in 1790, Hunter surveyed that island. He returned to England, arriving in 1792, and his naval career continued. Hunter was appointed Governor of New South Wales in 1794, and assumed office the following year. Like Phillip, he had virtually autocratic power. He faced major problems with the *New South Wales Corps, was recalled, and handed over office to King in 1800. He then resumed his naval career, later being promoted Vice-Admiral (1810). He continued to give advice on New South Wales, and gave evidence to the Select Committee on Transportation (1812). He died in London. Hunter published *An Historical Journal of the Transaction at Port Jackson and Norfolk Island* (1793) and *Governor Hunter's Remarks on the Causes of the Colonial Expense of the Establishment of New South Wales* (1802).

HURLEY, JAMES FRANCIS (1885–1962), photographer and film maker, was born in Glebe, *Sydney, the son of a printer and trade union official and his wife, and educated at Glebe Public School. After working in a steel mill at Lithgow, he became a photographer in Sydney, where he held his first exhibition. He was the official photographer to Douglas *Mawson's Australasian Antarctic Expedition from 1911 to 1913. After filming Francis Birtles's car expedition in northern Australia, he returned to the Antarctic as official photographer to Sir Ernest Shackleton's Imperial Trans-Antarctic Expedition from 1914 to 1916, during which time he took his famous photographs of the *Endurance* breaking up in pack ice. As Captain Frank Hurley, he served as an official photographer with the first AIF in France, Belgium, and the Middle East in 1917 and 1918. He accompanied Ross and Keith *Smith on the final leg of their famous flight from England to Australia in December 1919. He undertook two expeditions to film in the Torres Strait Islands and Papua between 1920 and 1923, but came into conflict with Sir Hubert *Murray. After being refused entry to Papua in 1925, he made two films on Thursday Island and in Dutch New Guinea. He made two more visits to the Antarctic, to join Mawson's British–Australian–New Zealand Antarctic Research Expedition, between 1929 and 1931. He worked in Australia as Cinesound Studio's chief cameraman from 1932 to 1936. He was an official photographer with the second AIF in the Middle East, where he remained and made documentary films for the British Government from 1942 to 1946. Hurley, who influenced many other Australian film makers, then concentrated on still photography after returning to Australia. He died in Sydney. His films include *Home of the Blizzard* (1913), *In the Grip of Polar Ice* (1917), *The Ross Smith Flight* (1920), *Pearls and Savages* (1921), *Jungle Women* (1926), and *Southward Ho with Mawson* (1930). His photographic publications include *The Blue Mountains and Jenolan Caves* (1952), *Sydney from the Sky* (1952), and *Western Australia: A Camera Study* (1953).

HYDRO-ELECTRIC COMMISSION OF TASMANIA Statutory authority responsible for the generation, distribution and sale of electricity in Tasmania, the

HYDRO-ELECTRIC COMMISSION OF TASMANIA

*Advertisement for the **Hydro-Electric Commission of Tasmania**, entitled 'The Return of Prosperity', from the Illustrated Tasmanian Mail, 11 November 1931. (Archives of Tasmania)*

state with the largest hydro-electric system in Australia. It was established in 1930, replacing the Hydro-Electric Department, which had been created when the Tasmanian Government acquired the partly completed Great Lake scheme in 1914. Hydro-electric schemes have been developed in the Derwent River catchment area; the Great Lake area; the Gordon River catchment area, involving the controversial destruction of Lake Pedder; the Mersey-Forth catchment area; and the Pieman River catchment area.

IMMIGRATION

IMMIGRATION Australia's first immigrants, the ancestors of the *Aborigines, are believed to have arrived more than 40 000 years ago. *Transportation of convicts, mainly from Britain, provided the next group of immigrants, who arrived, with the *First Fleet, in 1788; the last arrived in 1868. Free settlers began to arrive in 1793, but did not come in sizeable numbers until the 1830s, when significant assistance schemes were introduced. (Various such schemes have continued intermittently throughout the nineteenth and twentieth centuries.) *Systematic colonization, a theory propounded by *Wakefield and others, had some influence on policies during the 1830s

*Some of the poignancy associated with **immigration** is captured in the painting* An Emigrant's Thoughts of Home *(1859) by Marshall Claxton (1812–81), an English-born artist who lived in Sydney from 1850 to 1854. (National Gallery of Victoria)*

and 1840s. Caroline *Chisholm and others also assisted immigrants. The *gold-rushes of the 1850s attracted many immigrants, including some Chinese people, upon whose entry restrictions were soon placed. Such restrictions foreshadowed the *White Australia Policy, which excluded non-Europeans from Australia and was pursued for a large part of the twentieth century. *Blackbirding brought some Pacific Islanders to Australia from the 1860s onwards. Immigration virtually ceased during the *Depression of the 1890s, the First World War, and the *Great Depression. During the first part of the twentieth century, especially the 1920s, much encouragement was given to British immigrants, through organizations such as the *Big Brother Movement, and government policies such as *'Men, Money, and Markets'. Some non-British immigrants arrived during the nineteenth century, and increasing numbers arrived in the period between the two world wars, but the first large-scale non-British immigration did not occur until after the Second World War, initially as a result of *Calwell's introduction of an extensive immigration scheme. Many immigrants worked on projects such as the *Snowy Mountains Hydro-Electric Scheme. Some 4.2 million immigrants, of whom about 60 per cent were assisted, arrived between 1945 and the mid-1980s. About 40 per cent of these came from Britain and Ireland, the remainder from Italy, Greece, Yugoslavia, the Netherlands, Germany, New Zealand, Poland, the USA, Vietnam, Turkey, Lebanon and elsewhere.

INDUSTRIAL WORKERS OF THE WORLD (IWW) A syndicalist organization, founded in Chicago in 1905 (syndicalism being a 'movement among industrial workers having as its object the transfer of the means of production from their present owners to unions of workers for the benefit of the workers'). Members of the IWW were also known as 'Wobblies'. The first Australian branch began in Sydney in 1907, and others followed. Australian members, who numbered about two thousand, were very active during the First World War, taking a strong stand against war and *conscription. Tom Barker, the editor of the IWW's journal *Direct Action*, was convicted several times for making statements prejudicial to recruiting, and was jailed. The organization became the focus of controversy when a series of fires occurred in Sydney in 1916. Twelve members were charged with conspiring to commit arson, to secure Barker's release unlawfully, and to excite sedition. All were found guilty on at least one count, and their sentences ranged from five to fifteen years. An appeal in 1917 led to the quashing of two convictions on the second count, and corresponding reductions in sentences. The Street Inquiry of 1918 confirmed the guilt of the accused. The Ewing Royal Commission of 1920 found that six of the twelve had been wrongfully convicted; that four had been rightfully convicted, but had served adequate sentences; that one's sentence should be reduced; and that the other's sentence was confirmed. Ten were released soon after in 1920, one served his full sentence before being released in 1921, and the other was released early in 1921. The organization was banned in 1916, and was virtually dissolved in Australia by the end of 1917.

INGAMELLS, REX *Jindyworobaks

Portrait of Isaac Isaacs, Australia's first native-born governor-general, taken at about the time of the Federal Convention of 1897–8. (Australian Natives' Association)

IRIAN JAYA A province of Indonesia, comprising the western part of the island of *New Guinea, which the Dutch had claimed in 1828, and nearby islands. It was known in Australia as Dutch New Guinea or West New Guinea. The Dutch transferred authority for it to the United Nations in 1962, and it became part of Indonesia in 1963. For a time it was known as West Irian.

IRRIGATION The 'action of supplying land with water by means of channels or streams'. *Deakin was an early supporter of irrigation for Australia. The *Chaffey brothers were largely responsible for establishing Australia's first major irrigation settlements, at Renmark and Mildura, during the late 1880s. Further schemes in various states soon followed. The *Snowy Mountains Hydro-Electric Scheme, begun after the Second World War, provides water for irrigation as well as power. By the 1970s large areas were under irrigation in Victoria (mainly along the Murray River) and New South Wales (mainly along the Murray and Murrumbidgee Rivers), sizeable areas in Queensland, and smaller areas in South Australia (also mostly along the Murray) and Western Australia (notably in the Kimberley area, as a result of the *Ord River Irrigation Scheme); a small area was under irrigation in Tasmania. Irrigation schemes have often proved to be uneconomic, and are considered to have increased salinity in the Murray River.

ISAACS, SIR ISAAC ALFRED (1855–1948), lawyer, politician and governor-general, was born in Melbourne, educated at schools in Yackandandah and Beechworth, and was a teacher from 1870 to 1875. He was a clerk in the Crown Law Department from 1875 to 1881, studied law at the University of Melbourne, was admitted as a barrister in 1880, and was appointed Queen's Counsel in 1899. Isaacs was a Member of the Victorian Legislative Assembly (1892–1901), served as Solicitor-General once (1893), and Attorney-General twice (1894–9, 1900–1), and was a member of the *Federal Convention (1897–8). He then entered Federal Parliament. He was a Protectionist Member of the House of Representatives from 1901 to 1906, and was Attorney-General from 1905 to 1906. He was appointed to the *High Court in 1906, becoming Chief Justice in 1930. Isaacs generally favoured the extension of the Commonwealth's powers (in contrast with Sir Samuel *Griffith, the Chief Justice until 1919, who favoured the defence of states' rights). The court's decision in the *Engineers' Case* (1920) reflected a move towards Isaacs's position. He was appointed a Privy Counsellor in 1924. His appointment as the first Australian-born governor-general of the Commonwealth of Australia early in 1931 was the subject of controversy. He held that position until 1936.

IWW *Industrial Workers of the World

J

JACKAROO An Australian slang term that came into use during the nineteenth century and was applied to inexperienced colonists, many of whom came from Britain. It often was, and is, used to describe young men gaining experience on station properties in order to become station-managers. It may have originated as a portmanteau of Jack and kangaroo. The term jillaroo, to describe a female station-hand, was later adapted from it.

JANSSEN (or Jansz), **WILLEM** (c. 1570–?), Dutch mariner, is the first European known to have discovered Australia.

A dreadful fate befalls an inexperienced jackaroo.

He had sailed from Holland for the Dutch East Indies (present-day Indonesia), in command of the *Duyfken*, in 1603. Janssen charted part of the eastern coast of the Gulf of Carpentaria, which he thought was part of New Guinea, early in 1606. One of his crew was killed in an altercation with Aborigines. Janssen commanded several other vessels in the Dutch East Indies, before becoming the Governor of Fort Henricus, on Solor, in 1614. While returning to the Dutch East Indies in 1618, as supercargo on the *Mauritius*, following a visit to Holland, he landed again in Australia, near present-day North West Cape. Janssen was appointed to the Council of the Indies in 1619, and fought against the British for some time. As vice-admiral, later admiral, he served in a British–Dutch fleet against the Philippines. From 1623 to 1627 he was the Governor of Banda (present-day Banda Aceh). After commanding an expedition to Persia (present-day Iran), he assisted in the defence of Batavia (present-day Jakarta). Janssen arrived back in Holland in 1629, where he reported on affairs in the Indies.

JINDYWOROBAKS The name given to members of an Australian nationalist literary movement. The poet Rex Ingamells (1913–55) founded the Jindyworobak club in Adelaide in 1938. The title was taken from an Aboriginal term meaning 'to annex' or 'to join'. The Jindyworobaks, who included poets such as Ian Mudie, Roland Robinson, and William Hart-Smith, wanted to develop an Australian culture, free from colonialism and related to the Australian landscape, in other words, 'to join' Australians to their environment. In

1938 Ingamells described Jindyworobaks as 'those individuals who are endeavouring to free Australian art from whatever alien influences trammel it, that is to bring it into proper contact with its material'. Aboriginal culture, D. H. Lawrence's *Kangaroo* (1923), and P. R. ('Inky') Stephensen's *The Foundations of Culture in Australia* (1936) were all strong influences on the Jindyworobaks. Some of the movement's ideas and work are to be found in Ingamell's *Conditional Culture* (1938) and *Jindyworobak Review (1938–48)* (1948), and in the Jindyworobak poetry anthologies which were published annually between 1938 and 1953. The *Angry Penguins arose to some extent in opposition to the Jindyworobaks. The Jindyworobak movement had largely faded away by the mid-1950s, after the Second World War had helped to end Australia's isolationism.

JONES, DAVID (1793–1873), merchant, was born in Wales, where he became a grocer's apprentice, and later worked in a general store. He also worked in London for some time, before arriving in Sydney in 1835 as the partner of a business man named Charles Appleton. After the partnership ended in 1838, Jones traded from a building on the corner of George Street and Barrack Lane. He retired in 1856, but returned to the firm in 1860 when it faced difficulties, and retired again in 1868. One of his sons followed him into the business. David Jones was also a Member of the New South Wales Legislative Council from 1856 to 1860. He died in Sydney.

K

KALGOORLIE (originally Kalgurli) is a mining centre situated about 600 kilometres east of Perth. After Patrick ('Paddy') Hannan discovered gold there in 1893, a rush followed, and Kalgoorlie became the service centre for the nearby Golden Mile, one of the world's richest gold reefs. Gold production in the area peaked in about 1902. The population rose to about 200 000 in 1905, but was about 20 000 in the 1980s. Kalgoorlie was linked by rail with Perth in 1896, and with Esperance in 1927. In 1895 the Western Australian Premier, *Forrest, announced the decision to construct the Gold-fields Water Supply, which ensured a permanent water-supply to the area; it was opened in 1903. Charles Yelverton O'Connor, the engineer responsible for its construction, committed suicide in 1902, apparently because of criticisms of the scheme. Although it has declined in latter years, gold production in the area continued in the 1980s. Nickel was found near Kalgoorlie during the 1960s, and it has overtaken gold in importance.

KALGOORLIE RIOTS Riots in the Western Australian mining towns of Kalgoorlie and nearby Boulder in January 1934. The catalyst for the riots was the death of Edward Jordan early on 29 January, following a brawl with Claudio Mattaboni, a barman from Gianetti's Home from Home Hotel. Native-born Australians destroyed, damaged, and stole property belonging to southern Europeans, predominantly Italians and Yugoslavs, in Kalgoorlie and Boulder. On 30 January some miners decided to strike unless assured that unnaturalized 'foreigners' would not be employed in the mines. On the same day Australians attacked houses and camps, mostly occupied by Yugoslavs, at Dingbat Flat, near the Boulder railway station. In the fighting, Charlie Stokes, an Australian, was stabbed to death, and Joseph Katich, a Montenegrin, was shot and killed. During the two days about seventy buildings (including hotels, clubs, boarding-houses, cafés, shops, and houses), and a similar number of camps, were destroyed or damaged. Extra police arrived from Perth on 31 January and 2 February, but no further riots occurred after 30 January. A public appeal provided some relief for the victims. About ninety people were charged with various offences, mostly relating to looting. Most were fined; a few received short jail sentences. Mattaboni, who had been charged with manslaughter, was acquitted, and open findings were recorded regarding the deaths of Stokes and Katich. Australian resentment towards the southern Europeans was based partly upon allegations that the latter caused accidents in the mines (because of lack of knowledge of English and lack of experience), and that they paid 'slingbacks' (bribes) in return for jobs. Extreme heat and excessive drinking of alcohol were also factors leading to the riots.

KANAKAS *Blackbirding

KANGAROO ISLAND lies about 15 kilometres off the coast of South Australia. It is about 145 kilometres long, and 55 kilometres wide (at its widest point). *Flinders was perhaps the first White person to discover it, in 1802. *Sealers used the island in the following years. South Australia's first permanent White settlers, organized by the *South Australian Company, landed there in 1836.

KEATING, PAUL JOHN (1944–), politician, was born at Bankstown, New

KEATING, PAUL JOHN

*Paul **Keating**, Australia's Prime Minister from 1991, as seen by Spooner in his 1989 cartoon* High Interest Rates, *drawn when Keating was the federal Treasurer. (La Trobe Collection, State Library of Victoria)*

South Wales, into an Irish Catholic family. His father was a boiler-maker. Keating attended De La Salle College, Bankstown, until shortly before his fifteenth birthday, and later evening classes at Belmont and Sydney Technical Schools. He worked as a clerk at the Sydney County Council for some years, then for a Hong Kong trading company, before becoming a research officer and industrial advocate for the Federated Municipal and Shire Council Employees' Union in 1968. The following year he entered Federal Parliament as an *Australian Labor Party Member for Blaxland, New South Wales, in the House of Representatives. Keating was the Minister for Northern Australia for several weeks in 1975, before the defeat of the *Whitlam Government; president of the New South Wales Labor Party from 1979 to 1983; and Treasurer from

1983 to 1991 in the *Hawke Government, before replacing Hawke as Prime Minister in 1991.

KELLY, EDWARD (1855–80), *bushranger, was born in Victoria. His parents were Irish, his father an ex-convict. Ned Kelly had had several altercations with the law before the outbreak of the Kelly gang. Fitzpatrick, a police trooper, claimed that Ned had shot him when he attempted to arrest Ned's brother Dan in 1878. As a result of this incident, Ned's mother and two other people served jail sentences. Rewards of £100 each were offered for Ned and Dan, who hid in the Wombat Ranges near Mansfield. Joseph (Joe) Byrne and Stephen (Steve) Hart joined them. The four became known as the Kelly gang. The outbreak consisted of four main events, which took place at Stringybark Creek, Euroa, Jerilderie and Glenrowan, between 1878 and 1880. Ned shot and killed three policemen, Michael Kennedy, Thomas Lonigan, and Michael Scanlon, who were camped at Stringybark Creek while searching for the gang; a fourth policeman, Thomas McIntyre, escaped. Rewards of £500 for each member of the gang were offered. The gang then took over a station near Euroa. Byrne guarded the captives while the others held up a bank in Euroa, escaping with about £2000. Rewards were increased to £1000 each. Next the gang took over the police station at Jerilderie, New South Wales. Dressed in police uniforms, the gang held up the bank in Jerilderie, escaping with another £2000. They also took over the hotel. Rewards were then £2000 each. Some time after this, Byrne shot and killed Aaron Sherritt, a friend of his who had become a police agent. The gang's final stand was at Glenrowan, where they took over the hotel. Policemen, led by Francis Hare, surrounded the hotel. The members of the gang were dressed in armour made from plough mould boards. During the shoot-out Hare was shot in the arm, Byrne was shot and killed, Kelly was shot several times and was eventually captured, and two bystanders were killed. A policeman set

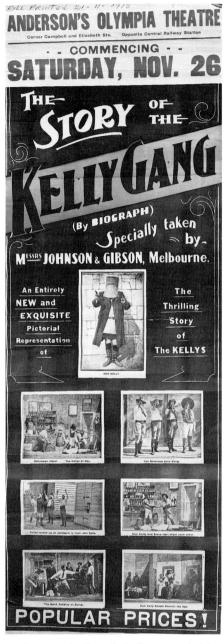

Advertising poster, printed on 21 November 1911, for The Story of the Kelly Gang *(1906). The film, which was re-released with additional scenes in 1910, presented a very sympathetic view of the exploits of Ned Kelly and his gang. It was one of a number of films about bushrangers made at that time. (National Library of Australia)*

fire to the hotel. Later, the badly burnt bodies of Dan Kelly and Hart were found; they may have committed suicide. Ned Kelly was found guilty of Lonigan's murder, was sentenced to death, and was hanged on 11 November 1880. Before his capture he had written two explanations of his actions; they are known as the 'Cameron Letter' and the 'Jerilderie Letter'. Kelly has become a folk hero. His experiences have inspired stories, ballads, books, paintings, films, and other works.

KIDMAN, SIR SIDNEY (1857–1953), pastoralist, was born near Adelaide. His father, a farmer, died shortly after his birth. Kidman was educated at private schools at Norwood. He worked variously as a stockman, carter, and butcher in New South Wales, established coaching businesses in New South Wales and Western Australia, and bought and sold cattle and horses. He acquired a series of stations, eventually owning or controlling land greater in size than England. Kidman, the 'Cattle King', died in Adelaide.

KING, PHILIP GIDLEY (1758–1808), *governor, was born in England, and pursued a naval career. He sailed for New South Wales with the *First Fleet. *Phillip chose King to establish a settlement on *Norfolk Island, which he did in 1788 and supervised that settlement until 1796. He briefly visited England to report upon the progress of New South Wales (1790–1), and was appointed Lieutenant-Governor in 1791. King returned to England because of ill health in 1796. After his health improved, he returned to New South Wales and relieved *Hunter in 1800. King promoted farming by small settlers in the colony, encouraged the development of whaling and sealing, and reduced liquor imports (but was unable to put an end to illegal distilling of spirits). Like Hunter, he faced major problems with the *New South Wales Corps. His governorship ended in 1806. He was in very poor health when he sailed for England in 1807, and he died the following year. One of his sons, Phillip Parker King (1791–1856), did a considerable amount of work charting parts of the Australian coastline.

KING ISLAND, an island measuring about 65 kilometres by 25 kilometres, lies in Bass Strait to the north-west of Tasmania, of which it has been a municipality since 1908. William Campbell sighted the island in 1797, John Black named it after Governor *King in 1801, John Murray charted its coast in early 1802, and both *Baudin and Charles Grimes visited it in late 1802. *Sealers and *whalers used the island in the early nineteenth century, and a few settlers followed them, but permanent settlement did not begin until 1888. Tom Farrell discovered scheelite (one of the tungsten ores) at King Island in 1904. Scheelite mining began during the First World War, and has continued intermittently since then; tin and other minerals have also been mined at various times. *Soldier settlement took place after each of the two world wars. The island's pastoral industry has continued throughout the twentieth century. King Island has a notorious reputation for shipwrecks.

KINGSFORD SMITH, SIR CHARLES EDWARD (1897–1935), aviator, was born in Brisbane, and educated in Canada and Sydney. During the First World War he served with the first AIF, transferred to the Australian Flying Corps

*Unsigned and undated portrait of Philip Gidley **King**, Governor of New South Wales from 1800 to 1806. (Mitchell Library, State Library of New South Wales)*

*Charles **Kingsford Smith** (left) and Charles Ulm in 1928, after making the first flight across the Pacific. (La Trobe Collection, State Library of Victoria)*

in 1916, and joined the British Royal Flying Corps in 1917. After returning to Australia in 1921, he flew joy-flights and later mail services, and established various businesses involving flying. 'Smithy', as he became known, created numerous records, and broke many others. His flights included a round-Australia flight with Charles Ulm, in record time in 1927; the first crossing of the Pacific (backed by Sidney *Myer), with Ulm and two Americans in 1928; the first crossing of the Tasman, with Ulm and two Australians in 1928; the first east-west crossing of the Atlantic in 1930; a solo flight from England to Darwin, in record time in 1930; and the first east-west crossing of the Pacific, with P. G. Taylor in 1934. Kingsford Smith and J. T. Pethybridge disappeared while on a flight from England to Australia. They are believed to have crashed near Burma on or about 8 November 1935. Kingsford Smith's publications include *The Old Bus* (1932), *My Flying Life* (1937), and (with Ulm) *Story of 'Southern Cross' Trans-Pacific Flight* (1928).

KINGSTON, CHARLES CAMERON (1850–1908), politician, was born in Adelaide. His father was an engineer (and later a politician). Kingston was educated at the Adelaide Educational Institution, and practised as a barrister; he was admitted to the Bar in 1873, and was appointed Queen's Counsel in 1888. He was a Member of the House of Assembly from 1881 to 1900, and a Member of the Legislative Council briefly in 1900. Kingston was Premier from 1893 to 1899, and held various

*Charles **Kingston**, Premier of South Australia from 1893 to 1899, was a leading member of the federation movement. (Australian Natives' Association)*

other portfolios. His Government introduced women's *suffrage, passed arbitration and conciliation legislation, and established a state bank. Kingston played a prominent role in the *federation movement. In 1901 he entered federal politics, and was a Protectionist Member of the House of Representatives until his death. He was Minister for Trade and Customs from 1901 to 1903.

KIRKPATRICK, JOHN SIMPSON (1892–1915), soldier, was born in Shields, County Durham, England, the son of a merchant seaman and his wife. After attending local schools he worked as a milk boy for four years, then spent a year with the 4th Durham Territorials, before leaving England with the merchant navy in 1909. The following year he deserted at *Newcastle, New South Wales. Jack (as he called himself) Kirkpatrick then worked variously on the Queensland cane-fields, the New South Wales coal-mines, the Western Australian gold-fields, and on ships around the Australian coast. Using the name John Simpson, he joined the first AIF in Perth at the beginning of the *First World War. He served as a private with the 3rd Field Ambulance, Australian Army Medical Corps, landing on Gallipoli on 25 April 1915. He and his donkey carried the wounded from the front to the medical stations, until he was shot and killed at Monash Valley on 19 May 1915. The 'Man with the Donkey', who was mentioned in dispatches but received no other honours or awards, became a legend, symbolizing bravery and self-sacrifice, commemorated in stories, stamps, paintings, and statues, the best-known examples of the latter being those at the *Australian War Memorial in Canberra and the Shrine of Remembrance in Melbourne.

KISCH AFFAIR A controversial diplomatic incident in 1934–5. Egon Erwin Kisch, a Czechoslovakian communist writer, was invited to speak at a Melbourne congress, organized by the Australian branch of the Movement Against War and Fascism in 1934. The *Lyons Government banned Kisch from entering Australia. He was not allowed to leave his ship in Fremantle, but jumped ashore at Port Melbourne (breaking his leg in the process), and was taken to Sydney where he failed an immigration test (a dictation test administered in Scottish Gaelic). (Kisch spoke many other languages.) The case was taken to court, but the *High Court found that Scottish Gaelic was not a 'European language' as required by the *Immigration Act*, and therefore the test was invalid. Kisch spoke at a number of meetings before leaving Australia in 1935. His publications include *Australian Landfall* (1937).

KOKODA TRAIL A foot-track that crosses the Owen Stanley Range in Papua New Guinea, beginning near Port Moresby in the south and ending near the small village of Kokoda in the north. It is about 240 kilometres long. During the Second World War Australian and Japanese soldiers fought along the Kokoda Trail for some months in 1942, until the latter were ordered to withdraw. It was the closest that any enemy forces have come to Australia by land.

KOREAN WAR (1950–3) Popular name given to the 'police action' that followed the North Korean attack on South Korea on 25 June 1950. The South

KOREAN WAR

Tom V. Carter's drawing of Simpson (more properly John Simpson **Kirkpatrick**) and his donkey, An Errand of Mercy. *(The Victorian Readers)*

Koreans were supported by *United Nations Organization forces, made up of representatives from Australia, Belgium, Canada, Colombia, Ethiopia, France, Greece, Luxembourg, the Netherlands, New Zealand, the Philippines, Thailand, Turkey, the Union of South Africa, the UK, and the USA (which led the United Nations intervention). The armistice concluded at Panmunjom on 27 July 1953 divided Korea along the 38th Parallel into North and South. This was similar to the position before the fighting. Within days of the North Korean attack, Australia offered support to South Korea, and sent members of the *Royal Australian Navy, the *Royal Australian Air Force, and the Royal Australian Regiment. There were about 1400 Australian casualties (including 278 deaths), most of them members of the Royal Australian Regiment.

L

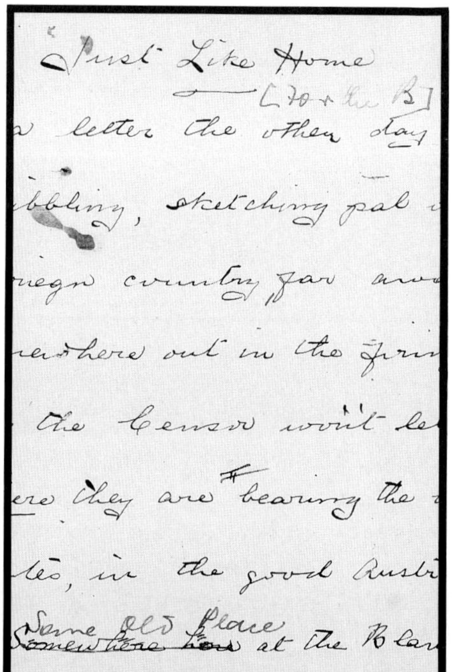

"Just Like Home
— [to + the B]
a letter the other day
...bbling, sketching pal
...reign country far aw...
...ewhere out in the Fir...
...the Censor won't le...
...ere they are bearing the
...ties, in the good Austr...
Same old Place
~~Somewhere~~ at the B...

LABOR PARTY *Australian Labor Party

LABOUR DAY *Eight-hour day movement

LAKE MUNGO An ancient lake, dry for the last 15 000 years, which is situated in south-western New South Wales, some 150 kilometres north-west of Balranald. Important archaeological discoveries providing evidence of early Aboriginal life have been made in the area since 1968. They include Mungo Woman, who was cremated about 26 000 years ago (the earliest known example of cremation in the world), Mungo Man, whose ochre-covered body was buried about 30 000 years ago, and tools and shells from about 33 000 years ago. Both Mungo Woman and Mungo Man are of gracile (light) build.

LALOR, PETER (1827–89), leader of the *Eureka Rebellion, and politician, was born and educated in Ireland. Before migrating to Australia in 1852 he was a civil engineer. He worked on the construction of the Melbourne–Geelong railway, before going to the gold-fields, firstly the Ovens, and then Ballarat, where he became involved in the Ballarat Reform League. In 1854 Lalor led the *diggers in the Eureka Rebellion, during which he was wounded, losing an arm as a result. He escaped after the Rebellion, and a reward of £200 was offered for information leading to his arrest, but was later withdrawn. The public raised money for Lalor, with which he bought land. He entered Victorian politics, and was a Member of the Legislative Council (1855–6), then a Member of the Legislative Assembly (1856–71, 1874–89). He held various offices, including that of Speaker (1880–7). During his political career he was also the director of several mining companies. Lalor died in Melbourne.

*Peter **Lalor** compares the results of different stages of his career. (La Trobe Collection, State Library of Victoria)*

LAMBERT, GEORGE WASHINGTON THOMAS (1873–1930), artist, was born in St Petersburg, Russia. His father, who died before Lambert's birth, was an American railway engineer; his mother was English. The family lived in Germany for some time, and in England (where Lambert was educated), before settling in New South Wales in 1887. Lambert worked variously as a clerk, a station hand and a grocer's assistant. He studied art under Julian Ashton in Sydney, began to contribute illustrations to the *Bulletin in 1895, and developed a reputation as a painter in the latter half of the 1890s. After winning the New South Wales Society of Artists' first travelling fellowship, he left Australia in 1900, studied art in Paris for about a year, and then settled in London, where he continued to draw and paint. He became known in particular for his portraits. In 1917 Lambert became an official Australian war artist, a position he held for some years. He settled in Sydney again in 1921. The art critic Alan McCulloch claims that between 1920 and 1930 Lambert's work 'dominated the Australian scene'. Lambert also executed a number of sculptures during the 1920s. He died at Cobbity, New South Wales. His paintings include *Across the Black Soil Plains* (1899), *Anzac: The Landing* (1918–22), and *The Charge of the Light Horse at Beersheba* (1920).

LAMBING FLAT RIOTS A number of violent clashes in 1860–1 between Europeans and Chinese *diggers on the Burrangong gold-fields (present-day Young), New South Wales. Lambing Flat was one of the areas where the Chinese were camped. The clashes, also known as the Burrangong Riots, culminated in a confrontation between the European diggers and the authorities on 14 July 1861. Similar riots had occurred elsewhere, notably at Buckland River. At Burrangong, 'roll-ups' of diggers assaulted the Chinese, forced them to leave, and destroyed their property. The main clashes occurred in November and December 1860, and January, February, and June 1861. A Miners' Protection (or Protective) League, whose aims included the exclusion of Chinese from the field, was formed at Burrangong early in 1861. It played some part in the riots until its demise in mid-1861. Military forces were stationed at Burrangong from March to May 1861. The arrest of three diggers on 14 July 1861, on charges relating to the June riot, led to a confrontation with the authorities. About one thousand diggers demanded the release of the prisoners, their demand was refused, and a riot ensued. A digger, William Lupton, was killed; three policemen and about twenty diggers were injured. The commissioners and police withdrew to Yass the following day. Further military and naval forces arrived in Burrangong at the end of July, and some remained there for a year. Several men were arrested on charges relating to the Lambing Flat Riots. Only two were convicted: Claremont Owen and William Spicer were each sentenced to two years' imprisonment. The New South Wales Government paid some financial compensation to the Chinese diggers. Earlier attempts to pass restrictive legislation in New South Wales had failed, but after the Lambing Flat Riots the *Chinese Immigration Restriction and Regulation Act 1861* was passed. It was repealed in 1867.

LAND RIGHTS A term that, in an Australian context, usually refers to

Aboriginal political and property rights. White settlers established missions and reserves for *Aborigines from the early nineteenth century onwards, but did not consider that Aborigines had legal rights to these or other lands. The land rights movement began in the 1960s, although some earlier claims had been made. There has been, and continues to be, opposition to it. In 1963 Yirrkala Aborigines from Arnhem Land presented a bark petition to the Commonwealth Parliament, protesting over a government decision to excise part of their reserve for mining. A Select Committee inquired into the protests, but mining went ahead. Some Yirrkala Aborigines later sought compensation, but in 1971 Mr Justice Blackburn held that they did not have legal rights to the land. Aborigines at *Wave Hill station were among others who played a significant role in the early stages of the land rights movement. Aboriginal land trusts, with varying powers, were established in South Australia in 1966, Victoria in 1970, Western Australia in 1972, and New South Wales in 1974. In 1972 Aborigines set up a tent embassy for about six months on the lawns of Parliament House in Canberra to publicize the land rights movement. The Aboriginal Land Rights Commission, under Mr Justice Woodward, was held in 1973 and 1974, and was followed by the passing of the *Aboriginal Land Rights (Northern Territory) Act* 1976, which led to the granting of some land rights to Aborigines in the Northern Territory. In 1975 an Aboriginal Land Fund Commission was established to administer funds to buy land throughout Australia for Aborigines. (An Aboriginal Development Commission took over its work in 1980.) In 1981 the South Australian Government, following extensive negotiations, granted the Pitjantjatjara people freehold title to some 100 000 square kilometres of land. There have been other developments since in various parts of Australia.

LANE, WILLIAM (1861–1917), journalist and trade unionist, was born in England, educated at Bristol Grammar School, and worked in various jobs, including that of journalist, in Canada and the USA. He migrated to Brisbane in 1885, where he became prominent in journalism and the trade union movement. Lane helped to found the *Boomerang* in 1887, the *Worker* in 1890, and the Australian Labour Federation in 1899. He was involved in the *Maritime and *Shearers' Strikes of 1890 and 1891, then led *New Australia, and later Cosme, the two Australian Utopian settlements in Paraguay. After leaving Cosme in 1899, Lane briefly edited the *Australian Worker* in Sydney in 1900 before settling in Auckland. He was leader writer of the *New Zealand Herald* from 1900 to 1913, then editor until his death. Lane had become conservative, founding the National Defence League in 1906, and arguing, among other things, for compulsory military training. His publications include *The Workingman's Paradise: An Australian Labour Novel* (1892), published under the pseudonym 'John Miller', and *Selections from the Writings of Tohunga* (1917), a collection of essays reprinted from the *New Zealand Herald*.

LANG, JOHN DUNMORE (1799–1878), clergyman, was born in Scotland, and migrated to Australia in 1823. Sydney's first Presbyterian minister, Lang played a prominent and often controversial role in colonial affairs. He was responsible for the building of Scots Church in

William Lane, the founder of New Australia. (Archives of Business and Labour, Australian National University)

LANG, JOHN THOMAS

Sydney (completed in 1826), broke away from the official Presbyterian Church in 1842 and was not formally reinstated for more than twenty years, and became the Moderator of the General Assembly of the Presbyterian Church in New South Wales in 1872. He promoted education, arranged various emigration schemes from Britain, founded and ran several newspapers, including the *Colonist* (1835–40), the *Colonial Observer* (1841–4), and the *Press* (1851), and wrote many books and pamphlets. Lang, who held radical and republican views, served on the Legislative Council of New South Wales several times during the 1840s and 1850s, and argued for the separation of the *Port Phillip District and the *Moreton Bay District (both of which he represented at different times on the Council) from New South Wales. His publications include *An Historical and Statistical Account of New South Wales, both as a Penal Settlement and as a British Colony* (1834), which is his best-known book, *Transportation and Colonisation* (1837), and *Freedom and Independence for the Golden Lands of Australia* (1852).

J. T. (Jack) Lang, often known as the 'Big Fella', was twice Premier of New South Wales. (The Herald & Weekly Times Ltd)

LANG, JOHN THOMAS (1876–1975), politician, was born in Sydney, and held various jobs, including that of estate agent, before becoming a Labor Member of the New South Wales Legislative Assembly in 1913. He was Treasurer three times (1920–2, 1925–7, 1930–2), Premier twice (1925–7, 1930–2), led the Opposition three times (1923–5, 1927–30, 1932–9), and also held other offices. His first Government restored the 44-hour week, introduced widows' pensions and child endowment, and extended workers' compensation. Lang tried unsuccessfully to abolish the Legislative Council in 1926 and 1930. During the *Great Depression his so-called Lang Plan, proposed in 1931, was controversial. It involved the suspension of interest payments to British bond holders, the reduction of interest on Australian government borrowings to 3 per cent, and the introduction of a new form of currency. His attempts to deal with the depression led to splits within the Labor Party, and his dismissal from the premiership by Sir Philip Game, the Governor of New South Wales, in May 1932 (followed by his party's electoral defeat a week later). Lang was expelled from the Labor Party in 1943 (but readmitted in 1971), resigned from the New South Wales Parliament in 1946, then served as an Independent Labor Member of the House of Representatives from 1946 to 1949, when he lost his seat. In 1938 he founded the *Century*, a weekly newspaper, and continued to edit it for the rest of his life. His publications include *Why I Fight* (1934), *The Great Bust* (1962), and *The Turbulent Years* (1970).

LA PÉROUSE, JEAN-FRANÇOIS DE GALAUP, COMTE DE (1741–88), was a French naval officer. In 1785 he sailed from France in charge of an expedition to explore the Pacific Ocean, his ships being the *Boussole* and the *Astrolabe*. He sailed around Cape Horn to Chile, then to Alaska via the Sandwich Islands (present-day Hawaii), south along the coast to California, across to China and the Philippines, north to Sakhalin and Kamchatka, and then south to Samoa, where islanders killed the *Astrolabe*'s commander and eleven other men. La Perouse then sailed to *Botany Bay via *Norfolk Island. On 26 January 1788 he entered Botany Bay, camped on the north shore, and met some members of the recently arrived *First Fleet. His

This cartoon, entitled The Youthful Larrikin, *from Harry Furniss's* Australian Sketches, *gives an indication of the clothing typically worn by* **larrikins**. *(National Library of Australia)*

expedition disappeared after leaving there in March, apparently wrecked soon afterwards in the Santa Cruz Islands. Its fate was not known until the 1820s.

LARRIKINS were the Australian equivalent of hoodlums or hooligans, rowdy persons who roamed the streets of Melbourne and Sydney in 'pushes' during the latter part of the nineteenth century. They wore distinctive clothing, and their activities ranged from rowdiness and pranks to rape and murder. The term seems to have originated in Melbourne.

LASSETER'S REEF An apparently mythical gold reef in Central Australia. In 1929 and 1930 Lewis Hubert (Harold Bell) Lasseter claimed, in various conflicting versions of the story, that many years before he had found a huge gold reef. In 1930 a seven-person expedition, led by Fred Blakeley and guided by Lasseter, left Alice Springs to search for the reef. No sign of it was found. Although the rest of the party abandoned the search, Lasseter continued. He died, probably of starvation, in the Petermann Ranges. In 1931 a search party found and buried his body and recovered his diary, in which he claimed to have rediscovered the reef. Ion Idriess's novel *Lasseter's Last Ride* (1931), and Blakeley's account, *Dream Millions* (1972), have helped to keep the myth alive. There have been several other unsuccessful attempts to find Lasseter's Reef.

LATHAM, SIR JOHN GREIG (1877–1964), lawyer, was born in Melbourne, was educated at Scotch College and the University of Melbourne, and taught for some time before studying law. While studying, and shortly after, he tutored and wrote for newspapers. He was admitted to the Bar in 1904. During the First World War he worked in naval intelligence. He accompanied the Australian delegation to the *Paris Peace Conference. Latham, who was appointed King's Counsel in 1922, was a Member of the House of Representatives from 1922 to 1934, first as an Independent, then representing the *Nationalist Party, and finally the *United Australia Party. He held various portfolios at different times, and led the Nationalist Party and the Opposition from 1929 to 1931. He was appointed to the Privy Council in 1933. Latham returned to the Bar in 1934, and was appointed Chief Justice of the *High Court in 1935, a position which he held until 1952. He was considered to be a 'centralist' in his interpretation of the Constitution. During the Second World War he was Australia's first Minister to Japan (1940–1). His publications include *Australia and the British Commonwealth* (1929).

*Sir John **Latham**, Chief Justice of the High Court from 1935 to 1952.*

LA TROBE, CHARLES JOSEPH (1801–75), colonial administrator, was born in London. He taught, travelled, and wrote several books about his travels. In the late 1830s La Trobe wrote three reports on education in the West Indies for the British Government. In 1839 he became Superintendent of the newly settled *Port Phillip District. La Trobe supported separation from New South Wales (although he was not active in the separation movement), but some Port Phillip colonists considered that he was too subservient to Sydney, and an unsuccessful petition requesting his removal was sent to England. He became Lieutenant-Governor in 1851 when the area was separated and renamed Victoria. Almost immediately he was faced with the problem of handling the *goldrushes. He adopted the New South Wales licence system, which later became a cause of the *Eureka Rebellion. He was responsible for setting aside considerable areas of parkland around Melbourne. La Trobe's term of office ended in 1854, and he left the colony that year. His publications include *The Alpenstock: or Sketches of Swiss Scenery and Manners* (1829), *The Rambler in North America, 1832–1833* (1835), and *The Solace of Song* (1837).

LAUNCESTON, a city in northern Tasmania, is situated at the junction where the North and South Esk Rivers form the Tamar River, a tidal estuary that runs into Bass Strait at Port Dalrymple some seventy kilometres from Launceston. *Bass and *Flinders discovered Port Dalrymple in 1798, and Hamelin, commanding one of *Baudin's ships, visited it in 1802. The following year, David *Collins sent William Collins from Port Phillip to explore the area. In 1804 William *Paterson established a settlement at York Town (near the mouth of the Tamar River on the west side) comprising mainly soldiers and convicts, having first camped at the site of George Town (on the east side of the

Tamar). He founded a settlement at Launceston (briefly known as Patersonia, but then renamed after *King's birthplace) in 1806, and it soon became the chief settlement in the north of Van Diemen's Land. *Macquarie's attempts to develop George Town in this role, rather than Launceston, failed. Launceston was declared a town in 1826, was made a municipality in 1852, and became a city in 1888. In 1876 it was connected by rail to *Hobart, about 200 kilometres to the south. Hydro-electric power has assisted its development. Launceston is Tasmania's second-largest city, after Hobart.

LAWSON, HENRY (1867–1922), poet and short-story writer, was born in New South Wales. His Norwegian father (Niels Larsen) had been a seaman, was a *digger at the time of Henry's birth, and later became a *selector. His mother was Louisa *Lawson. Henry Lawson was educated at schools at Eurunderee and Mudgee, and at night-school in Sydney. He was deaf by the age of fourteen. Lawson held numerous jobs, including those of farm worker, apprentice coach painter, clerk, house painter, sawmill worker, telegraph linesman, and teacher, in Australia and New Zealand; alcoholism was a problem for much of his life. He wrote for various newspapers and journals, including the Sydney *Bulletin, the Sydney Republican, the Brisbane Boomerang, and the Sydney Worker. Lawson visited London between 1900 and 1902; he died in Sydney. Lawson is one of Australia's best-known writers, and much of his work is about the bush and bush workers. His poetry includes *In the Days When the World Was Wide and Other Verses* (1896), *When I Was King and Other Verses* (1905), *The Skyline

Sir Francis Grant's Charles Joseph La Trobe (1855–6). La Trobe was Superintendent of the Port Phillip District from 1839 to 1851, then Lieutenant-Governor of Victoria from 1851 to 1854. (La Trobe Collection, State Library of Victoria)

Riders and Other Verses* (1910), *Song of the Dardanelles and Other Verses* (1916), *Too Old to Rat* (1917), and *Joseph's Dreams* (1923). His many other publications include *Short Stories in Prose and Verse* (1894), *On the Track* (1900), *Joe Wilson and His Mates* (1901), *The Romance of the Swag* (1907), and *Triangles of Life and Other Stories* (1913).

LAWSON (née Albury), LOUISA (1848–1920), feminist, journalist and social reformer, was born in New South Wales. Her father was a labourer. She married Niels Hertzberg (Peter) Larsen (later known as Lawson) in 1866, and bore several children, including Henry *Lawson. She worked at various jobs in

LAWSON, WILLIAM

Frederick Strange's streetscape Brisbane Street, Launceston 1858 *was painted six years after* **Launceston** *became a municipality. Strange, a former convict, specialized in such works. (Queen Victoria Museum and Art Gallery, Launceston)*

the country, including keeping a store and post-office. In the early 1880s Louisa Lawson left her husband and moved to Sydney. In 1887 she founded the *Republican* (later the *Australian Nationalist*), a short-lived journal, and the following year *Dawn*, which is said to have been Australia's first feminist journal. It advocated women's *suffrage and other rights. Louisa Lawson edited and published *Dawn* until its demise in 1905. She founded the Dawn Club, a women's social reform club, in 1889. She also published books, including Henry Lawson's first, *Short Stories in Prose and Verse* (1894), and some of her own stories and poems. She died in Sydney. Her own publications include *The Lonely Crossing and Other Poems* (1905).

LAWSON, WILLIAM *Blue Mountains, Crossing of the

LEAGUE OF NATIONS An international organization established under the Treaty of *Versailles after the First World War. Australia was one of the fifty-eight founding members. The League's main aims were to solve international disputes peacefully, and to foster international co-operation. Its ability to do either was hampered by the fact that the USA was never a member, and some members withdrew: Brazil in 1926, Japan in 1933, Germany (which had joined in 1926) in 1933, Italy in 1937, and Russia (a member from 1934) in 1940. Its system of sanctions failed. The League's final meeting was in 1946, shortly after the end of the Second World War. It was superseded by the *United Nations Organization.

LEICHHARDT, FRIEDRICH WILHELM LUDWIG (1813–48?), naturalist and explorer, was born in Prussia. His father was a farmer. Leichhardt was educated at the universities of Berlin and Göttingen, but did not take out a degree. After spending some years studying and making field trips in England, France, Italy and Switzerland, he arrived in Sydney in 1842. He led an overland expedition from the *Moreton Bay district to *Port Essington in the Northern Territory in 1844–5. He and six members of his party (two others having turned back, and one other having been killed by Aborigines) completed the

LEICHHARDT, FRIEDRICH WILHELM LUDWIG

Manuscript of Henry Lawson's 'Just Like Home'. (National Library of Australia)

journey in about fourteen and a half months. Leichhardt was hailed as a hero, was later given awards by geographical societies in London and Paris, and was also later pardoned for being a Prussian military deserter. He planned another expedition, aiming to leave from the Darling Downs and travel across the north to the western coast of the continent, and then to travel south to the Swan River settlement. His first attempt began in 1846, but failed. The party, which consisted of eight including Leichhardt, returned to the Darling Downs in mid-1847. His second attempt began in March 1848. The party, consisting of seven including Leichhardt, was last seen in the Darling Downs in April 1848. At least nine major searches between 1852 and 1938 failed to discover the fate of Leichhardt and his party, although some bones and relics, possibly belonging to them, were found. Leichhardt's publications include *Journal of an Overland Expedition in Australia, from Moreton Bay to Port Essington* (1847) and numerous articles.

LEND-LEASE Name given to American aid, such as munitions, materials, and services, to neutral and Allied countries during the Second World War. Congress passed the *Lend-Lease Act* 1941 while the USA was still neutral. After it became a belligerent, Reciprocal Lend-Lease was introduced. Britain and the USA made a Mutual Aid Agreement in 1942; Australia made similar agreements with the USA in 1942 and Canada in 1944. Lend-Lease finished in 1945, after the war ended. In 1946 Australia agreed to pay the USA US$27 million (of which US$20 million had already been paid). It was estimated that Australia received Lend-Lease worth approximately £A 466 million, and contributed Reciprocal Lend-Lease worth approximately £A 285 million to the USA.

LEONSKI, EDWARD JOSEPH (1917–42), murderer, was born in New York. He worked in various clerical jobs, before being conscripted into the United States Army in 1941. While stationed at Royal Park (temporarily renamed Camp Pell), *Melbourne, during the *Second World War, Leonski committed the so-called 'Brown-out Murders', named because they occurred during the brown-out (a period of restricted lighting due to wartime conditions). He strangled Ivy Violet McLeod at Albert Park on 3 May 1942, Pauline Buchan Thompson in Spring Street on 9 May, and Gladys Hosking at Royal Park on 18 May. The United States Army court-martialled Leonski. After unsuccessfully pleading not guilty, he was hanged at Melbourne's Pentridge Prison on 9 November 1942. The treatment of Leonski's case was unusual. Normally he would have been tried for such crimes in an Australian civil court; instead he was court-martialled by American military authorities.

LEWIS, ESSINGTON (1881–1961), industrialist, was born in Burra, South Australia. He was educated at St Peter's College, Adelaide, and the South Australian School of Mines, and became a mining engineer. Lewis joined the *Broken Hill Proprietary Company Limited (BHP) and was to play a leading role in its development. During the First World War he managed the Broken Hill Munitions Company Proprietary Limited from 1915 to 1918. He became BHP's assistant general manager in 1918, general manager in 1921, managing director in 1926, and chief general manager in 1938. During the Second World War he held various public offices, including those of Director-General of Munitions, from 1940 to 1945, and Director-General of Aircraft Production from 1942 to 1945. After the war Lewis continued as chief general manager of BHP until 1950, chaired the company from 1950 until 1952, and then remained on its board until he died. He wrote several publications about Australia's iron and steel industry.

LIBERAL PARTY An Australian political party. The *Fusion Party had generally become known as the Liberal Party by 1913, when *Deakin resigned as the federal parliamentary leader of the Liberal Party, and from Commonwealth Parliament. Joseph *Cook replaced him as leader. The Liberals governed federally from June 1913 to September 1914, when they were defeated by Labor. After the Labor Party split over conscription, the Liberals merged with the *National Labor Party to form the *Nationalist Party in 1917. Liberals have also governed, sometimes in coalition with other parties, in various states at various times. The Nationalist Party was succeeded in turn by the *United Australia Party and the *Liberal Party of Australia.

LIBERAL PARTY OF AUSTRALIA Australia's major non-Labor political party. It superseded the *United Australia Party in 1945. The Liberal Party governed federally, in coalition with the Country Party (*National Party of Australia), from 1949 to 1972, and again from 1975 to 1983. *Menzies, who was largely responsible for founding the Liberal Party, was its federal parliamentary leader from 1945 to 1966. He was followed by *Holt (1966–7), *Gorton (1968–71), *McMahon (1971–2), Billy Mackie Snedden (1972–5), *Fraser (1975–83), Andrew Sharp Peacock (1983–5, 1989–90), John Winston Howard (1985–9), and John R. Hewson (1990–). Liberals have also governed, sometimes in coalition with other parties, in various states at various times.

LIGHT, WILLIAM (1786–1839), soldier and surveyor, was born in Malaya and educated in England. Before becoming the first surveyor-general of the new British province of South Australia in 1836, Light's varied experiences included two years' service in the British navy, a military career (during part of which he served in the Peninsular War), and extensive travel. He arrived in South Australia in 1836, chose the site for Adelaide, and began surveys of it. *Hindmarsh later actively opposed the choice of site. Light also began surveys elsewhere in South Australia, but was hampered by insufficient staff and equipment, and his requests to England for more of both were rejected. After being instructed to undertake temporary surveys for the sake of speed, he resigned in 1838. He then became the main partner in a private surveying firm. He died the following year. His publications include *A Brief Journal of the Proceedings of William Light* (1839).

'**LIGHT ON THE HILL**' *Chifley, Joseph Benedict

LINDSAY FAMILY A well-known Australian artistic and literary family. Its best-known member is NORMAN ALFRED WILLIAM LINDSAY (1879–1969), artist and writer. He was born in Victoria. His father was an Irish-born medical practitioner. Norman Lindsay was educated at Creswick Grammar School, then worked as an illustrator in Melbourne for some time. In 1901 he moved to Sydney, where he became an illustrator for the *Bulletin*, a position he held for many years. His prolific artistic output includes cartoons, line drawings, water-colours, oil paintings, etchings, and sculptures. His best-known art works are probably those drawings and paintings that are, in Bernard Smith's words, 'dominated by deep-bosomed and

LINDSAY FAMILY

*Norman Lindsay, the best-known member of the famous **Lindsay family**, is remembered as both a writer and an artist. (Mitchell Library, State Library of New South Wales)*

heavy-hipped nudes luxuriating in exotic and bacchanalian settings'. His literary output was also prolific. Both his art and writing were controversial. Norman Lindsay, who had travelled in Europe and America, died at Springwood in the Blue Mountains, where he had lived for some fifty years. His many novels include *Redheap* (1930), which was banned in Australia for many years, *Saturdee* (1933), and *Halfway to Anywhere* (1947). One of his children's books, *The Magic Pudding* (1918), which he wrote and illustrated, has become a classic. Some of his personal philosophies are discussed in *Creative Effort* (1920), *Hyperborea* (1928), and *Madame Life's Lovers* (1929). His *Bohemians of the Bulletin* (1965), *The Scribblings of an Idle Mind* (1966), and *My Mask* (1970) are autobiographical. Norman Lindsay also wrote articles for periodicals such as *Vision*. Other members of the family include the painter and illustrator SIR DARYL ERNEST LINDSAY (1890–1976), one of Norman's brothers, who was for many years director of the National Gallery of Victoria; LADY JOAN À BECKETT LINDSAY (née Weigall) (1896–1984), Daryl's wife, water-colourist and writer, whose publications include *Time Without Clocks* (1962) and *Picnic at Hanging Rock* (1967); JOHN (JACK) LINDSAY (1900), Norman's son, writer and translator; SIR LIONEL ARTHUR LINDSAY (1874–1961), another of Norman's brothers, graphic artist, water-colourist and writer, whose publications include *The Australian Work of Arthur Streeton* (1919), *Conrad Martens* (1920), *The Art of Hans Heysen* (1920), and *Charles Kean* (1934); PERCIVAL (PERCY) CHARLES LINDSAY (1870–1952), another of Norman's brothers, water-colourist and illustrator; and RUBY LINDSAY ('Ruby Lind') (1887–1919), Norman's sister, illustrator, who married the cartoonist Will *Dyson.

LOGAN, PATRICK *Moreton Bay

LONGFORD, RAYMOND HOLLIS (né John Walter) (1878–1959), film director, was born in Melbourne. He began acting in Australian films in about 1909, and began directing in about 1911. Longford, the best-known Australian silent-film director, directed numerous films (some of which he also acted in) during the next twenty years, including *The Romantic Story of Margaret Catchpole* (1911); *The Silence of Dean Maitland* (1914); *The Mutiny of the Bounty* (1916); *The Sentimental Bloke* (1919), often described as an Australian film classic; *Ginger Mick* (1920); *On Our Selection* (1920); and *Fisher's Ghost* (1924). Longford and Lottie Lyell, who starred in many of the films, formed a partnership to produce some of them; Lyell also co-directed at least one. Longford continued to work in the film industry in various minor capacities during the 1930s. He spent his last years as a night watchman on the wharves in Sydney, where he died.

'LUCKY COUNTRY' Popular term used to describe Australia, particularly the Australia of the 1960s. It came from Donald Horne's study of contemporary Australian society, *The Lucky Country* (1964), which included such comments as 'Australia is a lucky country run mainly by second-rate people who share its luck'. Many people mistakenly thought that the term was one of praise for Australia's prosperity and stability. As Horne wrote in 1970, 'a phrase that was intended as an ironic rebuke became a phrase of self-congratulation'.

LYELL, LOTTIE *Longford, Raymond Hollis

LYNE, SIR WILLIAM JOHN (1844–1913), politician, was born in Apslawn, Tasmania. His father was a politician. Lyne was educated by a tutor, and at Horton College, Ross, Tasmania; spent some time in the Gulf of Carpentaria area; was a council clerk in Glamorgan, Tasmania, from 1865 to 1875, when he became a pastoralist at Camberoona, near Albury. He entered politics in New South Wales, and was a Member of the Legislative Assembly from 1880 to 1901. During this time he was Premier from 1899 to 1901 and held various other offices. Hopetoun invited Lyne, who was a strong opponent of the *federation movement, to become the first prime minister of the Commonwealth of Australia; Lyne was unable to form a

LYONS, DAME ENID

Joe Lyons, Australia's Prime Minister from 1932 to 1939. (National Library of Australia)

ministry, and *Barton became Prime Minister instead. Lyne was a Member of the House of Representatives from 1901 until his defeat in 1913, shortly before his death. He was Minister for Home Affairs (1901–3), Minister for Trade and Customs (1903–4, 1905–7), and Treasurer (1907–8). Lyne was a Protectionist for most of his political career.

LYONS (née Burnell), DAME ENID MURIEL (1897–1981), was born in Tasmania, educated at Stowport and Burnie state schools and the Teachers Training College in Hobart, and taught for a year. She married Joseph *Lyons in 1915, and had six sons and six daughters. Dame Enid was a Member of the House of Representatives from 1943 to 1951, the first woman to hold that position. She represented the *United Australia Party, and then its successor, the *Liberal Party of Australia. She also became the first woman member of Federal Cabinet; she was the Vice-President of the Executive Council from 1949 to 1951. She was also active in various other organizations. Her publications include *So We Take Comfort* (1965), *The Old Haggis* (1969), and *Among the Carrion Crows* (1972).

LYONS, JOSEPH ALOYSIUS (1879–1939), politician, was born in Tasmania to Irish parents, was educated at a convent school in Ulverstone, Stanley State School, and the Teachers Training College in Hobart, and became a teacher. He was a Labor Member of the Tasmanian House of Assembly from 1909 to 1929. He was Treasurer (1914–16, 1923–8), Leader of the Opposition (1916–23, 1928–9), Premier (1923–8), and held other portfolios. His Government's achievements included improving the state's finances. Lyons was then a Member of the House of Representatives from 1929 until his death, representing Labor (1929–31), and then the *United Australia Party (1931–9). When *Theodore was reinstated as Treasurer in 1931 Lyons and others left the Labor Party, joined with Nationalists, and formed the United Australia Party. Lyons advocated a deflationary policy to deal with the *Great Depression, with reductions in public expenditure, wages, and pensions. He led the Opposition from 1931 to 6 January 1932, and then was Prime Minister until his death, leading a United Australia Party Government until 1934, and then one in coalition with the Country Party. He also held other portfolios. Dame Enid *Lyons was his wife.

M

*An illustration by Joseph Lycett of John **Macarthur**'s property, Elizabeth Farm, Parramatta, published in London in 1825. Macarthur and his wife Elizabeth did much to develop the Australian fine-wool industry. (National Library of Australia)*

MACARTHUR, JOHN (1767–1834), pastoralist, was born in England to Scottish parents. His father was a mercer and draper. Macarthur pursued a military career, arriving in New South Wales in 1790 as a lieutenant in the *New South Wales Corps. He became regimental paymaster at Parramatta in 1792, inspector of public works in 1793, and was promoted captain in 1795. Macarthur's conflicts with authorities eventually earned him his nickname, 'the Perturbator'. He clashed with *Hunter and *King, who sent him to England after he wounded William *Paterson in a duel in 1801. Macarthur resigned from the army before returning to New South Wales to pursue his mercantile and pastoral interests in 1805. He played an active role in the *Rum Rebellion of January 1808, after which he became Colonial Secretary until July. In 1809 he sailed for England, and remained there for some years, aware that he risked arrest if he returned to New South Wales. When he did return to the colony in 1817, it was on the condition that he did not take part in public affairs. He came into conflict with *Macquarie, and later with *Darling. *Brisbane's attempt, in 1822, to appoint him as a magistrate failed. Macarthur was a Member of the Legislative Council from 1825 to 1832. He played an active role, greatly assisted by his wife Elizabeth, in the development of the Australian fine wool industry; he publicized Australian fine wool during his visits to England; and he is said to have influenced the *Bigge reports. He took control of the *Australian Agricultural Company in 1828. Macarthur died in New South Wales.

McCubbin, Frederick (1855–1917), painter, was born in West Melbourne. His father was a baker. McCubbin studied at the Artisans School of Design, Carlton, and at the National Gallery School, Melbourne, where he taught drawing from 1886 until his death. He was president of the Australian Art Association once (1912), and of the Victorian Artists Society several times (1893, 1902–4, 1908–9, 1911–12). He visited Europe in 1907. McCubbin is best known for his landscapes, and was a leading member of the *Heidelberg School. According to Bernard Smith, 'Frederick McCubbin appears to have been the prime mover in promoting the distinctly national quality of the Heidelberg School'. His paintings include *The Lost Child* (1886), *Down on His Luck* (1889), *Bush Burial*, *Wallaby Track* (1896), and *The Pioneers* (1904).

McEwen, Sir John (1900–80), politician, was born at Chiltern, Victoria, and was educated at state schools, before working in the Crown Solicitor's Office in Melbourne from 1916 to 1918. He enlisted in the first AIF, but the First World War ended before he could be sent overseas. He became a soldier-settler, running a dairy farm at Stanhope in the Goulburn Valley, joined the Victorian Farmers' Union in 1919, and was active in establishing a co-operative dairy factory in his district. McEwen entered federal politics; he was a Victorian Member of the House of Representatives from 1934 to 1971, representing the Country Party. He held several portfolios, and was successful in developing *protection in the period from 1949 to 1967. McEwen was Deputy Leader of the Country Party (1943–58), and then Leader (1958–71). He was Deputy Prime Minister from 1958 to 1971. After *Holt's disappearance, he became Prime Minister as a short-term measure (December 1967–January 1968) until *Gorton was elected. He returned to farming after retiring from politics in 1971.

MacGregor, Sir William (1846–1919), colonial administrator, was born and educated in Scotland. He became assistant medical officer in the Seychelles in 1873, medical officer in Mauritius in 1874, and chief medical officer in Fiji in 1875. In each of these places he also held administrative positions, including that of Acting Governor of Fiji for two periods. MacGregor became the Administrator of British New Guinea in 1888. During his term of office, which lasted until 1898 (his title becoming Lieutenant-Governor in 1895), he established a native police force, organized missionary activity, encouraged European development, and explored extensively. He was Governor of Lagos from 1898 to 1902, Governor of Newfoundland from 1904, and Governor of Queensland from 1909 to 1914. In 1910 MacGregor became the first chancellor of the University of Queensland, which he had helped to found.

Machine-breakers   *'Swing' Rioters

McIlwraith, Sir Thomas (1835–1900), politician, was born and educated in Scotland. In 1854 he migrated to Victoria, where he worked firstly as a miner, then as an engineer. He began to make investments in Queensland, and eventually moved there permanently.

*Frederick **McCubbin**'s* Lost *(1886), one of his history paintings, has become an icon. The lost child has been a prominent theme in Australian art and literature. (National Gallery of Victoria)*

McIlwraith was a Member of the Queensland Legislative Assembly for many years (1870–1, 1873–86, 1888–96); he was Premier three times (1879–83, 1888, 1893), and held various other offices. He attempted unsuccessfully to annex south-eastern New Guinea on Britain's behalf in 1883. He died in London.

MCKAY, HUGH VICTOR (1865–1926), inventor and businessman, was born at Raywood, Victoria. His father was a farmer. McKay's harvesting machine, which stripped, threshed and cleaned grain, was patented in 1885. He established McKay's Harvesting Machine Company at Ballarat in 1891, but moved it to Braybrook (later renamed Sunshine), to the west of Melbourne, in 1906. His application for exemption from excise duties under *New Protection legislation led to the *Harvester Judgement of 1907. McKay's company reputedly was the largest manufacturer of agricultural machinery in the southern hemisphere by the 1920s.

MCKELL, SIR WILLIAM JOHN (1891–1985), governor-general, was born in New South Wales, and was educated at Bourke Street Public School, Surry Hills. He became a boiler maker, and was active in the Boilermakers' Union. McKell then entered state politics, and was a Labor Member of the New South Wales Legislative Assembly from 1917 to 1947. He held various portfolios, before becoming Premier and Treasurer (1941–7). He was Governor-General of the Commonwealth of Australia from 1947 to 1953.

MACKELLAR, ISOBEL MARION DOROTHEA (1885–1968), poet, was born in Sydney, the daughter of wealthy Australian-born parents. Her father was a physician. She was educated privately and at the University of Sydney, travelled widely in Australia and overseas, and became fluent in a number of languages. Her poem, 'Core of My Heart', was written in about 1904, and compared English and Australian landscapes. It was published in the London *Spectator* in September 1908, and reprinted as 'My Country' in the *Sydney Mail* in October 1908. It became very popular during the First World War and has been reprinted in many anthologies. The following lines, comprising the second of its six stanzas, are probably the best-known lines of Australian poetry:

> *I love a sunburnt country,*
> *A land of sweeping plains,*
> *Of ragged mountain ranges,*
> *Of droughts and flooding rains.*

The poem was included in Dorothea Mackellar's *The Closed Door* (1911). Her other volumes of verse include *The Witch-Maid* (1914), *Dreamharbour* (1923), and *Fancy Dress* (1926). She wrote one novel alone, *Outlaw's Luck* (1913), and two with Ruth Bedford, *The Little Blue Devil* (1912), and *Two's Company* (1914). She was awarded an OBE shortly before her death in Paddington.

MCKILLOP, MARY HELEN (1842–1909), Mother Mary of the Cross, was born in Melbourne and educated at private schools and by her father, who had studied for the priesthood, before working variously as a shop assistant, governess, teacher, and proprietor of a girls school in Portland. In March 1866 she and Father J. E. Tenison Woods founded a religious society named 'The Sisters of St Joseph of the Sacred Heart' in the South Australian town of Penola, with

Photograph of Dorothea Mackellar, c. 1915–16. By this time her poem 'My Country', expressing her love for the Australian landscape, had already become very popular. (Mitchell Library, State Library of New South Wales)

Portrait of Mother Mary McKillop, co-founder of the Sisterhood of St Joseph of the Sacred Heart. (Sisterhood of St Joseph of the Sacred Heart)

Mary as the first member and superior. The society, originally founded to educate poor children in remote areas, soon spread to other parts of South Australia and broadened its activities to include other forms of social work. After some priests became disaffected towards the sisters, Bishop Sheil excommunicated Mary in September 1871 for alleged insubordination and the order was partly disbanded. Sheil lifted the excommunication in February 1873 shortly before his death. Mary visited Rome and received papal approval for the order later that year, and travelled through Europe visiting schools before returning to Adelaide in January 1875. Mother Mary established schools, convents, and other institutions throughout Australia. Disaffection between priests and sisters again arose, an investigation was conducted into the sisterhood, and Sheil's successor, Bishop Reynolds, exiled Mother Mary from Adelaide. She then transferred the mother house, or headquarters, to Sydney. After suffering a stroke in May 1901, she remained an invalid until her death in Sydney. Several steps have been taken towards Mother Mary becoming Australia's first saint. In 1926 the cause was opened, but it was suspended in 1930 and not reopened until 1951. In 1973 she was formally declared a servant of god, and in 1991 a group of Rome theologians unanimously agreed that she was worthy of being declared venerable.

McMahon, Sir William (1908–88), politician, was born in Sydney, and educated at Sydney Grammar School and the University of Sydney. Before and after the Second World War he practised as a solicitor, and during the war he served with the Militia. He entered federal politics, and was a Liberal Member of the House of Representatives from 1949 to 1982. McMahon held various portfolios, and was Deputy Leader of the Liberal Party from 1966 to 1971, when he became Prime Minister. His Government was defeated at the 1972 elections. McMahon then resigned as Leader, Billy Mackie Snedden being elected in his place.

Maconochie (né M'Konochie), **Alexander** (1787–1860), penal reformer, was born in Edinburgh, Scotland. He pursued an active career in the navy from 1803 to 1815, and spent part of this period as a prisoner-of-war of the French. He helped to found the Royal Geographical Society in London in 1830, and became its first secretary. Maconochie then became the first professor of geography at the University of London in 1833. He went to Van Diemen's Land, where he was a secretary to Sir John *Franklin from 1836 to 1839. Maconochie was Superintendent of the *penal settlement at *Norfolk Island from 1840 to 1844. During this time he implemented his now well-known mark system. He did not believe that punishment should be cruel or vindictive: prisoners should earn their freedom by labour and good behaviour, measured by marks, rather than by time; they should pass through various stages, including one in which they worked as members of a group. After his recall, Maconochie returned to England in 1844. He governed a new prison at Birmingham from 1849 to 1851. His publications include *A Summary View of the Statistics and Existing Commerce of the Principal Shores of the Pacific Ocean, etc.* (1818), *Report on the State of Prison Discipline in Van Diemen's Land . . .* (1838), and the extremely influential *Crime and Punishment, The*

Mark System, framed to mix Persuasion with Punishment, and make their Effect improving, yet their Operation severe (1846).

MACQUARIE, LACHLAN (1762–1824), *governor, was born on the island of Ulva in the Inner Hebrides, Scotland, and pursued a military career, which took him to various parts of the world. He was appointed Governor of New South Wales following the *Rum Rebellion, sailed with the 73rd Regiment in 1809, and took office in 1810. He was the last governor of New South Wales with virtually autocratic powers. Convict numbers increased dramatically during his term of office, especially after the end of the Napoleonic Wars in 1815. Macquarie used many of the convicts to carry out an extensive programme of public works, which included the construction of roads; numerous buildings, many of which were designed by *Greenway; and a number of new towns. His policy towards convicts and *emancipists (such as *Redfern), generally considered to be a supportive one, led to conflict with *exclusives such as *Macarthur, *Marsden, and the Bent brothers. He made various attempts to improve public morality, and treated Aborigines with sympathy. Macquarie (who had been promoted brigadier-general in 1811) resigned in 1821, was relieved by Thomas *Brisbane, and left the colony for England in 1822. The *Bigge reports, the result of an inquiry held in the latter stages of his term of office, distressed Macquarie. He did not hold public office again after leaving New South Wales, and died in London.

MACQUARIE HARBOUR   An inlet, named after Governor *Macquarie, on the western coast of Tasmania. During the *transportation period a *penal settlement was established there, the first forty-five convicts being sent in late 1821. The main settlement was on Sarah Island, and another settlement was located on a nearby smaller island. Convicts sent to Macquarie Harbour, which gained a notorious reputation as a place of punishment, worked at timber-felling, building of vessels, and other pursuits. The settlement was abandoned in late 1833.

MAHOGANY SHIP   The name given to a ship's hull found by sealers in 1836 in the sand dunes near the present-day Victorian city of Warrnambool. There were numerous sightings of it during the nineteenth century. It was possibly made from mahogany (hence its name) or cedar. The last recorded sighting of it was in 1880. Numerous searches since then have failed to find it, but have uncovered some relics in the area. It may have been burnt or hidden by sand. Some people believe that the Mahogany Ship may have been a Portuguese vessel of the fifteenth or sixteenth century.

MAJOR'S LINE   A route from New South Wales to *Australia Felix, named after Major Thomas *Mitchell. The tracks he made on his return from Australia Felix to Sydney in 1836 were visible for many years, and *squatters from New South Wales followed the Major's Line in search of new pastures.

MALAYAN EMERGENCY   A period of conflict in 1948–60 between Malayan-Chinese communists and British and other forces in Malaya, which at the time was ruled by Britain. Australia, at Britain's request, sent members of the *Royal Australian Air Force (1950–8) and the Royal Australian Regiment

Unsigned and undated portrait of Lachlan **Macquarie**, *Governor of New South Wales from 1810 to 1821, during which time large numbers of convicts were transported to the colony. (Mitchell Library, State Library of New South Wales)*

'MAN WITH THE DONKEY'

Michael Leunig used the legend of the **Mahogany Ship** to comment upon the state of public affairs in Victoria in The Wreck published in the Age, 4 April 1992. Yet another search for the remains of the ship was underway at the time. (Age)

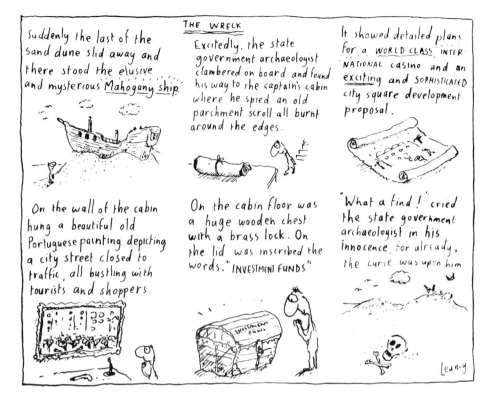

(1955–62) to Malaya. Eventually, the British defeated their opponents. The British tactics of fighting guerrillas during the Malayan Emergency were copied, with far less success, during the *Vietnam War.

'MAN WITH THE DONKEY' *Kirkpatrick, John Simpson

MANNIX, DANIEL PATRICK (1864–1963), Roman Catholic archbishop, was born in Ireland. His father was a tenant farmer. Mannix was educated at, taught at, and from 1903 to 1912 was president of, St Patrick's College, Maynooth, in Ireland. He was Coadjutor Archbishop (1913–17) and the Archbishop (1917–63) of Melbourne. Mannix became a controversial public figure. He argued for state aid to Catholic schools, opposed the British response to the Irish Easter Rising of 1916, argued for Irish independence, and spoke out strongly against *conscription during the referendum campaigns of 1916 and 1917. The British refused him entry to Ireland in 1920. Mannix became a strong opponent of communism, and was influential in the establishment of the National Secretariat of Catholic Action (1937) and the Catholic Social Movement, a major force in the creation of the *Democratic Labor Party. During the campaign for the 1958 federal elections he tried to discourage Catholics and others from voting for the Australian Labor Party. Mannix died in Melbourne.

MAORI WARS *New Zealand Wars

MARALINGA *Nuclear Tests

MARITIME STRIKE The name usually given to Australia's first almost-general strike, in 1890. It actually involved a series of strikes between August and November. Maritime workers were the first to strike, hence the popular but somewhat inaccurate name. Transport workers, miners, and pastoral workers followed. Some fifty thousand men in Victoria, New South Wales, Queensland, and South Australia became involved. Workers were fighting for the principles of unionism and the 'closed shop' ('an establishment in which only trade union members are employed', or 'a trade etc. restricted to members of a [particular] trade union'). Employers were fighting for the principle of 'freedom of contract' (the right to employ anyone). Other issues were also involved. The unions were defeated, and some collapsed completely.

MARK SYSTEM *Maconochie, Alexander

MARRIED WOMEN'S PROPERTY ACTS Popular name given to legislation that transferred rights over married women's property from the women's husbands to the women themselves. Such legislation was passed in the Australian colonies from the 1870s onwards.

MARSDEN, SAMUEL (1764–1838), chaplain, was born at Farsley, Yorkshire, England. He was apprenticed to his father, a blacksmith, and became a lay preacher. The Elland Society (an Anglican organization) sponsored his education at Hull Grammar School and Magdalene College, Cambridge. Marsden's studies were interrupted when he was appointed assistant chaplain of New South Wales. After ordination as a deacon and then as a priest, Marsden arrived in New South Wales in 1794. He visited England (1807–9), and was appointed senior chaplain in 1810. He played a prominent, and sometimes controversial, role in public affairs, and clashed with *Macquarie, especially over the latter's *emancipist policy. Marsden's public duties included that of magistrate, and he was active in organizations such as the London Missionary Society, the Church Missionary Society, and the British and Foreign Bible Society; in 1813 he founded the New South Wales Society for Affording Protection to the Natives of the South Sea Islands and Promoting their Civilization; he supervised missionaries in the Pacific Islands and New Zealand, and in the course of his missionary work visited New Zealand seven times between 1814 and 1837. Marsden was also a noted sheep-breeder. He died at Windsor, New South Wales.

MARSHALL, ALAN (1902–84), writer, was born at Noorat, Victoria, where his father was a storekeeper. Marshall was crippled with poliomyelitis in 1908. After attending school in Terang and business college in Melbourne, he worked in various jobs, including those of clerk, bookkeeper, and accountant. He began writing during the 1920s, and during the Second World War he wrote as a freelance journalist and then lectured for the Army Education Service. He travelled widely in Australia and overseas. Best known for his autobiographical *I Can Jump Puddles* (1955), which has been translated into many other languages and adapted for film and television, Marshall wrote numerous other stories, histories and travel books, and did much to record Australian folklore.

MARTENS, CONRAD (1807–78), artist, was born in London. His father, a German-born merchant, was for some time Austrian consul in London; his mother was English. He studied landscape painting under the water-colourist Copley Fielding. Martens left England in the *Hyacinth*, bound for India, but in 1833 joined the *Beagle* at Montevideo as an artist. While on board the *Beagle*, which was making a scientific survey of Patagonia and Tierra del Fuego, he became friendly with Charles Darwin. Martens left the *Beagle* at Valparaiso in 1834. After visiting Tahiti and New Zealand, he arrived in Sydney in 1835, and lived there for the rest of his life, undertaking various sketching expeditions and giving some art lessons. Martens is best known for his watercolour landscapes and seascapes, many of which are of Sydney Harbour and surrounding areas, although he also produced lithographs and other works. He was a parliamentary librarian from 1863 until his death. His paintings include *View from Neutral Bay* (c. 1857–8) and *Sydney from Vaucluse* (1864).

'MARVELLOUS MELBOURNE' A now-famous term which was, and is, used to describe Melbourne during the boom years of the 1880s, when it developed rapidly. The term was used because of the apparent material, commercial, social, and cultural progress of the city. George Augustus Sala, an English journalist touring the colonies, first used the term in newspaper articles published in the London *Daily Telegraph*, the Melbourne *Argus*, the Melbourne *Australasian*, and the *Sydney Morning Herald* in 1885. The term's use soon became widespread.

MAWSON, SIR DOUGLAS (1882–1958), geologist and Antarctic explorer, was born in England, and came to Australia as a child. He took part in a geological surveying expedition to the New Hebrides (present-day Vanuatu) in 1903. With degrees in engineering and science from the University of Sydney, he became a lecturer at the University of Adelaide in 1905. From 1907 to 1909 he took part in Ernest Shackleton's British Antarctic Expedition. Mawson and two others became the first people to reach the South Magnetic Pole, in 1909. He led the Australasian Antarctic Expedition (1911–14), aspects of which were recorded by Frank *Hurley. The expedition established three bases, mapped a large part of the Antarctic coast, and engaged in scientific research. While on a trip inland, Mawson's two companions died; he returned to base alone. Mawson was knighted in 1914. During the First World War he worked in the British Ministry of Munitions. He became Professor of Geology at the University of Adelaide in 1920, and from 1929 to 1931 he led the British–Australian–New Zealand Antarctic Research Expedition (BANZARE), and claimed British possession of part of Antarctica in 1930. Mawson continued to take an active interest in Australia's Antarctic involvement and research until his death. One of the bases in the *Australian Antarctic Territory was named after him. His publications include *The Home of the Blizzard* (1915).

MECHANICS INSTITUTES Organizations in Australia that, like their British predecessors, aimed to provide educational facilities for working-class adults (a mechanic being a 'skilled workman, esp. one who makes or uses machinery'). Official attempts to establish such an organization in Sydney in 1826 failed, but one was established by skilled tradesmen in Hobart in 1827, and others were established in Sydney, Adelaide,

Melbourne, and elsewhere in the 1830s. Such organizations, also known as schools of arts and by various other names, were eventually established in many towns and cities. They tended to attract middle-class members. Some were later absorbed by other educational institutions; others later became social rather than educational organizations.

MELBA (née Mitchell), DAME NELLIE (Helen Porter) (1861–1931), singer, was born in Melbourne. Her father was a building contractor. She was educated at the Presbyterian Ladies College, Melbourne, and studied singing under Pietro Cecchi, in Melbourne, from 1879. She married Charles Frederick Nesbitt Armstrong, the manager of a Queensland sugar plantation, in 1882, and bore a son the following year, but the marriage did not last. She went to Europe in 1886, studied singing under Madame Marchesi in Paris, and in 1887 made her operatic debut in Brussels, in *Rigoletto*. Melba, who had adopted her stage name in 1886, went on to sing in London, Paris, New York, and elsewhere, becoming the best-known soprano of the time. She toured Australia in 1902 and 1903. From 1909 onwards, she lived alternately in Australia and Europe. In 1911 and 1924 she brought Melba–Williamson Operas to Australia. On occasions she taught at Melbourne's Conservatorium of Music. She was created a Dame of the British Empire in 1918, and a Dame Grand Cross of the Order of the British Empire in 1927. Dame Nellie Melba gave a number of supposedly final performances: her final Covent Garden performance was in 1926; in 1927 she sang at the opening of Parliament House, Canberra; and her final Australian performance was in 1928. She died in Sydney. Her publications include *Melodies and Memories* (1925).

Portrait of Nellie Melba, the most famous soprano of her era. Photograph by Arnold Genthe, San Francisco, stamped 1901. (La Trobe Collection, State Library of Victoria)

MELBOURNE, the capital city of the state of *Victoria, lies on the Yarra River, near its mouth in Port Phillip Bay. Charles Grimes, Surveyor-General of New South Wales, explored the area in 1803. *Batman, a settler from Van Diemen's Land, followed in 1835, entered into a 'treaty' (later declared invalid) with local Aborigines to buy land, and allegedly noted 'the site for a village'. Later that year John Pascoe Fawkner's party, also from Van Diemen's Land, settled near Batman's site. *Bourke, the Governor of New South Wales, visited the settlement in 1837, named it Melbourne after the British Prime Minister, and arranged for its planning and first land sales. Melbourne became a municipality in 1842, and a city in 1847. Melbourne's population was approximately 283 000 in 1881, and 3 002 300 in 1988.

Collins Street at Four P.M., *from the* Illustrated Australian News, *18 July 1868, depicting one of* **Melbourne**'s *most fashionable streets during the prosperous years which followed the Victorian gold-rushes. (La Trobe Collection, State Library of Victoria)*

Melbourne*Voyager disaster

MELBOURNE CUP Australia's most famous horse-race. The first Melbourne Cup was run under the auspices of the Victoria Turf Club on Thursday, 7 November 1861. It has been run every year since then. The race was taken over by the Victorian Racing Club, which was founded in 1864. Originally it was run over 2 miles; since 1972 it has been run over 3200 metres. Notable winners have included Archer (1861, 1862), Carbine (1890) and *Phar Lap (1930). The Melbourne Cup is run at the Flemington racecourse on the first Tuesday in November, which is a public holiday in Melbourne.

'MEN, MONEY, AND MARKETS' A slogan used to sum up the theory behind many of the policies of the *Bruce–*Page Government (1923–9). 'Men' referred to immigrants, to increase Australia's population; 'Money' referred to capital, mostly to be borrowed from the British Government; 'Markets' referred to imperial preference, that is, mutual preference for trade between Great Britain and the dominions. During the 1920s Australia was more successful in attracting 'Men' and 'Money' than 'Markets'.

MENZIES, SIR ROBERT GORDON (1894–1978), politician, was born in Victoria, and educated at Wesley College and the University of Melbourne. His father was a shopkeeper and politician. Menzies became a barrister in 1918 and was appointed King's Counsel in 1929, was a Member of the Victorian Legislative Council in 1928–9, and a Member of the Victorian Legislative Assembly from 1929 to 1934, representing the Nationalists until 1931, and then their successor, the *United Australia Party. He held various offices. Menzies was then a Member of the House of Representatives from 1934 to 1966, during which time he represented the United Australia Party until 1944, and then the *Liberal Party of Australia, which he had been largely responsible for founding. Menzies held numerous portfolios, and was Prime Minister from 1939 to 1941, when he resigned because of dissatisfaction with his leadership; he led the Opposition from 1943 to 1949, and then was Prime Minister again from 1949 to 1966, when he resigned. He strongly supported close relations with Britain, hosted the first visit to Australia by a reigning monarch (Queen Elizabeth II) in 1954, and supported Britain's policy in the Suez Crisis of 1956. He supported the Dutch against Indonesia in the conflict over West New Guinea (*Irian Jaya). During his second term of office, Australian troops were sent to the *Korean War, the *ANZUS Pact and *SEATO treaty were signed, Australia became involved in the *Vietnam War, and the *Petrov affair occurred. Menzies was strongly anti-communist, and unsuccessfully attempted to outlaw the *Communist Party of Australia, first by legislation, and then by referendum. The Commonwealth Government's role in the provision of tertiary education expanded greatly during these years (particularly notable was the establishment of the Australian Universities Commission in 1959), as did its role in the field of secondary education. Canberra also developed considerably in this period. Menzies was Chancellor of the University of Melbourne from 1967 to 1972. His publications include *Afternoon Light* (1967) and *The Measure of the Years* (1970), both of which are memoirs.

MITCHELL, HELEN PORTER *Melba, Dame Nellie

The Cup Day: A Sketch on the Lawn, *from the* Australasian Sketcher, *29 November 1873. Melbourne society parading at Flemington race-course on* **Melbourne Cup** *Day, 6 November 1873. Don Juan won the race. (La Trobe Collection, State Library of Victoria)*

*Australia's longest-serving prime minister, Robert Gordon **Menzies** (shown here in a well-known photograph), held that position from 1939 to 1941 and again from 1949 to 1966. (National Library of Australia)*

MITCHELL, SIR THOMAS LIVINGSTONE (1792–1855), surveyor-general, was born in Scotland. He pursued a military career, during which he served in the Peninsular War. Mitchell, who had by then attained the rank of major, arrived in Sydney as Assistant Surveyor-General of New South Wales in 1827. He became Surveyor-General in 1828. His work in the colony included leading four major expeditions. On the first, in 1831–2, he travelled north-west from Sydney and explored an area around the Gwydir and Barwon Rivers. On the second, in 1835, he charted the Bogan River, and much of the Darling River. On the third, in 1836, he followed the Lachlan River, then the Murrumbidgee River to the Murray River, followed it to the Darling River, explored part of the latter, and then followed the Murray to the junction of the Loddon River. He then went south-west and discovered an area that he named *Australia Felix, reached the mouth of the Glenelg River and, several days later, Portland Bay (where he met *Henty), and then went in a north-easterly direction back to Sydney. *Squatters soon followed the *Major's Line to new pastures. On his fourth expedition, in 1845–6, he attempted to find an inland route to *Port Essington in the north, but failed, reaching an area near the site of present-day Isisford before returning to Sydney. He represented the *Port Phillip District on the Legislative Council of New South Wales in 1844–5. Mitchell, who had visited Britain several times since first coming to the colony, died in Sydney, while still in office. His publications include *Three Expeditions in the Interior of Eastern Australia* (1838), *Journal of an Expedition into the Interior of Tropical Australia* (1848), *The Australian Geography* (1850), and *Report upon the Progress Made in Roads and Public Works in New South Wales* (1856).

MOLESWORTH COMMITTEE A Select Committee of the British House of Commons (1837–8). It was appointed to inquire into *transportation and was chaired by Sir William Molesworth, hence its popular name. Its report criticized transportation in general, and assignment in particular. Shortly afterwards, the British Government abolished assignment in the Australian colonies, and suspended transportation to New South Wales. Many criticisms, some of which are debatable, have been made of the Molesworth Committee, notably that its members were prejudiced, that they were strongly influenced by Wakefieldian ideas, and that their use of evidence was questionable.

MONASH, SIR JOHN (1865–1931), engineer and soldier, was born in Melbourne. His father was a German-Jewish merchant. Monash was educated at Scotch College and the University of Melbourne. Before the First World War he practised as an engineer, becoming a pioneer in reinforced concrete construction. He joined the Melbourne University Rifles in 1884, was commissioned in the Garrison Artillery, commanded the Australian Intelligence Corps in Victoria in 1908, and became a colonel of an infantry brigade in 1913. During the First World War he took the 4th Australian Infantry Brigade to Egypt, and commanded it throughout the *Gallipoli campaign. He then took command of the 3rd Division, which he trained in England in 1916, and then took into action in France. His notable characteristics of meticulous planning, and lucid exposition of his plans (particularly to those who had to execute them), were already evident, and were exemplified again in his division's first major engagement, the Battle of Messines, which began on 7 June 1917. In May 1918 he took command of the Australian Corps. He further enhanced his reputation as a commander, particularly in the Battle of Amiens, which began on 8 August 1918. After the war Monash was appointed general manager of the Victorian State Electricity Commission in 1920, and in 1921 chairman of the Commission. He was responsible for the development of brown coal resources in the Latrobe Valley. Monash, who was also active in other public affairs, died in Melbourne. His publications include *The Australian Victories in France in 1918* (1920) and *War Letters* (1933). Monash University was named after him.

MONCRIEFF, GLADYS (1892–1976), singer, was born in Bundaberg, Queensland. She appeared in vaudeville, before being employed by *Williamson. 'Our Glad', as she became known, appeared in many operettas and musical comedies, including *The Maid of the Mountains* (in which she performed more than three thousand times between 1921 and 1952), *The Merry Widow*, *The Southern Maid*, *The Chocolate Soldier*, and *Rio Rita*. She appeared in London in the mid-1920s, in New Guinea during the Second World War, and in Korea during the *Korean War, but spent most of her life in Australia. Her autobiography, *My Life of Song*, was published in 1971.

MORAN, PATRICK FRANCIS (1830–1911), Roman Catholic archbishop, was born in Ireland and educated in Rome. He became a priest in 1853, and was vice-principal of St Agatha's College, Rome, from 1856 to 1866. Moran then returned to Ireland, where he was private secretary to Bishop Cullen, his uncle, from 1866 to 1872, and was Bishop of Ossory from 1872 to 1884, when he was appointed Archbishop of Sydney. In 1885 he was made a cardinal. Moran was responsible for building many Catholic institutions, including St Patrick's College (a seminary at Manly); supported the labour movement and federation; and advocated Home Rule. He wrote a number of books and pamphlets.

MORANT, HARRY HARBORD ('The Breaker') (1864–1902), poet and soldier, was born in England. After migrating to Australia in 1883, he worked in the bush for many years. It appears that, using the name Edwin Henry Murrant, he married Daisy O'Dwyer (later *Bates) at Charters Towers in Queensland in March 1884, but soon separated from her. Morant gained a reputation as a horse-breaker (hence his pseudonym) and as a poet, with some of his poetry being published in the *Bulletin*. After the *Boer War began, he enlisted in the South Australian Mounted Rifles and served in South Africa. After a brief period in England, Morant returned to South Africa in March 1901. He was commissioned as a lieutenant in the Bushveldt Carbineers, an irregular unit consisting largely of Australians. Following events in August and September 1901, Morant and several other Bushveldt Carbineers faced a British court martial on charges relating to the deaths of a number of Boer prisoners and a British missionary of German descent. Morant and an Australian named Peter Joseph Handcock were acquitted over the missionary's death, but were found guilty of some of the charges relating to the other deaths, and were executed, without the knowledge of the Australian

Government, by a firing squad at the Pretoria Gaol on 27 February 1902. Another Australian, George Ramsdale, was also sentenced to death, but his sentence was commuted to life imprisonment. After spending some years in an English gaol, he was released in 1904. An English irregular soldier named Picton was cashiered. The executions may have influenced the restrictions regarding death sentences which were included in the Australian *Defence Act 1903*. 'Breaker' Morant, who became a folk hero after his death, has been the subject of a film (1980), and several biographies.

MORETON BAY is a large bay on the Queensland coast near *Brisbane. It is flanked by Bribie, Moreton, and Stradbroke Islands. James Cook, who discovered it in 1770, named it after the Earl of Morton. John Oxley explored the area in 1823. A *penal settlement was established on its shores in 1824, but was soon moved to Brisbane, where it remained until 1839. Patrick Logan became the commandant of the settlement in 1826. He was killed by Aborigines in 1830, near present-day Mount Beppo, while on an exploratory expedition. Although free settlers were banned from the Moreton Bay District, some began to arrive in 1840, and the ban was lifted in 1842. The Moreton Bay District was separated from New South Wales and proclaimed the colony of Queensland in 1859.

MORRISON, GEORGE ERNEST ('Chinese') (1862–1920), doctor, journalist, and traveller, was born in Geelong, Victoria, the son of the principal of Geelong College and his wife. After attending his father's school, George Morrison began medical studies at the University of Melbourne, but failed an examination in March 1882. He had already written newspaper articles about his travels within Australia, and in 1882 he investigated the *blackbirding trade in Queensland. His six *Leader* articles, published between October and December that year, and subsequent critical letter to the *Age*, helped to end the practice. Morrison walked back to Melbourne from Normanton in the Gulf of Carpentaria, a distance of more than 3220 kilometres, between December 1882 and April 1883. In October 1883, while travelling in *New Guinea for the *Age* and *Sydney Morning Herald*, Morrison received several spear wounds. In 1884 he travelled to Edinburgh for treatment, and completed his medical studies there, graduating MB, Ch M, in August 1887. After travelling in North America and the West Indies, and working for eighteen months as a doctor in Spain, he returned to Australia in late 1890 and worked as resident surgeon at the Ballarat Base Hospital for two years from April 1891. In 1894 he walked across China to Burma, covering some 4830 kilometres in three months. After returning to Australia, he sailed to England, where his *An Australian in China, being the narrative of a quiet journey across China to Burma* (1895) was published and he graduated MD in August 1895. He returned to China and worked on a trial basis for *The Times*, before taking up an official position as that paper's correspondent in Peking (Beijing) in March 1897. Morrison was wounded during the *Boxer Rising in 1900, and reported on the Russo–Japanese War of 1904–5 and subsequent peace conference held at Portsmouth in the USA. In August 1912 he resigned from his position with *The Times* to become adviser to the new Chinese republican government. 'Chinese' Morrison, as he became known to Westerners because of his knowledge of China, advised the Chinese delegation to the *Paris Peace Conference, but left Paris in ill health for England in May 1919, and died in Devon a year later. Several biographies of Morrison have been written and Chinese residents in Australia inaugurated an annual memorial lecture in his name in Canberra in 1932.

MORSHEAD, SIR LESLIE JAMES (1889–1959), soldier and business man, was born at Ballarat, attended state schools and the Melbourne Teachers College, and worked as a schoolteacher for some years. Morshead, who became an officer in the 2nd Battalion of the AIF, was wounded during the *Gallipoli campaign. Having been promoted Lieutenant-Colonel, he commanded the 33rd Battalion from 1916 to 1919. His leadership on the Western Front was considered to be very successful. He was mentioned in dispatches six times during the First World War and received a number of awards. Between the wars he joined the Orient Line in 1924, became its Sydney manager in 1938, and served in the Militia. Morshead commanded the 18th Brigade of the second AIF from 1939 to 1941, and, as major-general, the 9th Division from 1941 to 1943. He played notable roles in the defence of Tobruk in 1941 and the Battles of El Alamein in 1942. From 1943 to 1945 in the Pacific, he commanded in turn the 2nd Corps, the New Guinea Force and the 1st Corps. He was mentioned in dispatches three times during the Second World War, and was made CBE in 1941, KBE in 1942, and KCB also in 1942. After the war he went back to the Orient Line and became its Australian manager in 1947, before retiring in 1954; he chaired a number of organizations, including the *Big Brother Movement; and headed the Committee of Inquiry into Defence Organization for the government in 1957–8.

MORT, THOMAS SUTCLIFFE (1816–78), business man, was born in Bolton, Lancashire. His father was a manufacturer. Mort arrived in Sydney as a clerk in 1838, became an auctioneer and broker in 1843, and ran Sydney's first regular wool sales. He became involved in various mining ventures; established a dock and engineering works at Balmain where ships and railway locomotives were built, and implemented a system of profit-sharing with employees for some time; and developed a large dairy estate at Bodalla, New South Wales. Mort contributed much to the development of refrigeration techniques, but died shortly before frozen meat was successfully shipped from Australia.

MOUNT ISA, one of Australia's major mining centres, lies in the Selwyn Range in north-western Queensland. John Campbell Miles, a prospector, discovered ore bearing silver and lead at Mount Isa, which he named, in 1923. Mount Isa Mines Limited was founded in 1924, soon took over leases, and developed the field, which has yielded copper, silver, lead, and zinc. The shire became a city in 1968. Mount Isa has been called the 'Copper City'.

MUELLER, BARON SIR FERDINAND JAKOB HEINRICH VON (1825–96), botanist, was born in Rostock, Schleswig-Holstein. His father was a customs commissioner. Mueller studied pharmacy and botany, before migrating in 1847 to South

Walter G. Mason's wood engraving T. S. Mort Esq., of Sydney. *Business man Thomas* **Mort** *is remembered particularly for his contribution to the development of techniques for the shipping of frozen meat. (Rex Nan Kivell Collection, National Library of Australia)*

*Baron Ferdinand von **Mueller**, shown here wearing numerous medals awarded in recognition of his work as a botanist, was the Victorian Government Botanist from 1853 until his death in 1896. (La Trobe Collection, State Library of Victoria)*

Australia, where he worked as a chemist and then as a farmer. He moved to Melbourne in 1852, was appointed Government Botanist in 1853 (a position which he held until his death), and undertook several expeditions in Victoria, Western Australia, and Tasmania, during which he noted many previously unknown plant species. He also took part in A. C. Gregory's North West Australia Expedition (1855–6). Mueller was the director of the Melbourne Botanical Gardens from 1857 to 1873, had extensive interests, belonged to numerous societies, and received many honours and awards. He published hundreds of papers and books, including *Fragmenta Phytographiae Australiae* (1858–82), *The Natural Capabilities of the Colony of Victoria* (1875), *Select Plants Readily Eligible for Industrial Culture or Naturalization in Victoria* (1876), *Eucalyptographia: A Descriptive Atlas of the Eucalypts of Australia and the adjoining Islands* (1879–84), and *Key to the System of Victorian Plants* (1885–8), and assisted with the preparation of George Bentham's *Flora Australiensis* (1863–78).

MUNRO-FERGUSON, SIR RONALD CRAUFURD *Novar, Ronald Craufurd Munro-Ferguson, Viscount

MURDOCH, SIR WALTER LOGIE FORBES (1874–1970), essayist, was born in Scotland, the son of a Presbyterian minister. He came to Victoria with his family in 1884, was educated at Scotch College and the University of Melbourne, became a schoolteacher, and also began writing for newspapers. Murdoch lectured in English at the University of Melbourne from 1903 to 1911, was Professor of English at the University of Western Australia from 1912 to 1939, and then Chancellor from 1943 to 1947. Western Australia's second university was named after him. His best-known works are the essays which he contributed to newspapers for many years. Some of these were later published in collections such as *Speaking Personally* (1930), *Saturday Mornings* (1931), and *Moreover* (1932). He also edited a number of anthologies, including *The Oxford Book of Australian Verse* (1918). His other publications include *Alfred Deakin: A Sketch* (1923).

MURRAY, SIR JOHN HUBERT PLUNKETT (1861–1940), administrator, was born in Manly, New South Wales, and educated in Australia and England, where he was called to the Bar from the Inner Temple in 1886. After returning to Australia, he practised as a barrister. During the Boer War he commanded the New South Wales Irish Rifles. He was appointed chief judicial officer of British New Guinea in 1904, became acting administrator of the area in 1907, and served as the first lieutenant-governor of Papua under Australian rule from 1908 to 1940. He attempted to encourage European enterprises in

Papua, did much to bring Papua under government control, and introduced welfare measures in fields such as education and health. Difficulties facing him included problems relating to the First World War, antagonism displayed by some White settlers (especially in the period around 1920), the effects of the *Great Depression, and a general lack of finance. Sir Hubert Murray died while in office.

MUSIC CONSERVATORIUMS have been founded in various places in Australia. One in Melbourne, attached to the University of Melbourne, was founded in 1894. Some of its functions have since been transferred to the Victorian College of the Arts, which first took music students in 1974. The Conservatorium became known as the University's Faculty of Music. The Elder Conservatorium of Music, attached to the University of Adelaide, was founded in 1898. The New South Wales State Conservatorium of Music (the first director of which was Henri Verbrugghen) was founded in 1916, and the Newcastle Conservatorium of Music in 1952. The Queensland Conservatorium of Music was established in 1957, and the Tasmanian Conservatorium of Music in 1962. The latter became part of the University of Tasmania in 1981, and that University's Faculty of Music in 1984. A number of other schools of

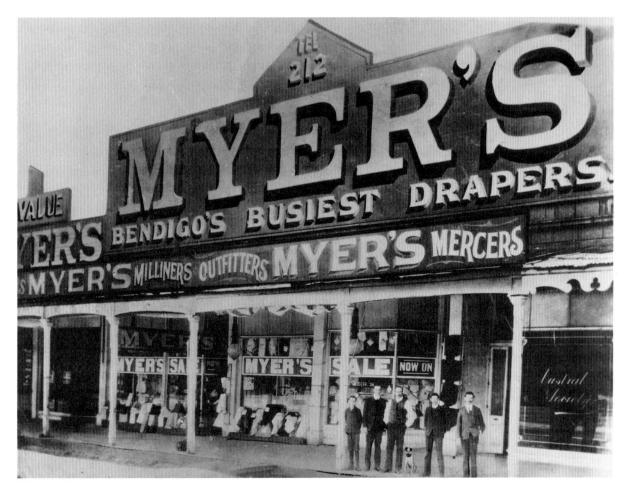

*One of Sidney **Myer**'s Bendigo drapery stores, clearly showing his flair for publicity. (Myer Stores Ltd)*

music and university departments of music have also been established.

MYALL CREEK (near present-day Inverell), New South Wales, was the scene of the killing of at least twenty-eight Aboriginal men, women and children, by White station hands in June 1838. As a result, eleven Whites were tried for murder, but were acquitted. However, seven of them were retried, found guilty, and hanged. There was much public protest: prosecution of Whites for murdering Aborigines was unusual.

'MY COUNTRY' *Mackellar, Isobel Marion Dorothea

MYER, SIMCHA (SIDNEY) BAEVSKI (1878–1934), business man and philanthropist, was born in Russia, the son of a Hebrew scholar, Ezekiel Baevski, and his wife. He joined his brother Elcon (1875–1938) in Melbourne in 1898. After adopting the family name Myer, they opened a drapery store in the Victorian town of *Bendigo later that year. Sidney bought out his brother soon after they moved to other premises in Bendigo in 1902. In 1908 Sidney, who showed considerable flair for advertising and display, bought another drapery store in the same town. Elcon later managed the two Bendigo stores before they were sold in 1914. Sidney bought yet another drapery store in Bourke Street, Melbourne, in 1911. The business flourished, and the store was rebuilt. The Myer Emporium, based on San Francisco's Emporium, was opened in July 1914. Sidney acquired other businesses, including woollen mills in Ballarat in 1918. A public company was formed in 1925, an Adelaide department store was bought and a Myer Emporium was established there in 1928, and by the early 1930s the stores sold a very wide range of goods. Sidney Myer donated money to many organizations and causes, including the University of Melbourne, *Kingsford Smith's air crossing of the Pacific in 1928, Melbourne's Children's Hospital, the Melbourne Symphony Orchestra, the building of the Yarra Boulevard during the Great Depression, and the Shrine of Remembrance, and left money to establish a Sidney Myer Trust. Other family members continued in the business after his death. In 1985 the Myer organization merged with the company founded by G. J. *Coles.

N

NAMATJIRA, ALBERT (1902–59), artist, was born at the Lutheran Hermannsburg Mission near Alice Springs in the Northern Territory. An Aborigine of the Aranda people, he was educated at the mission school. In the mid-1930s Rex Battarbee tutored him in the use of water-colours. Namatjira became famous for his water-colour landscapes of central Australia. He was granted Australian citizenship in 1957. He became the subject of controversy in 1958, when he was charged with supplying liquor to members of his people (who were not Australian citizens). He was found guilty and sentenced to six months' imprisonment (later commuted to three months'), and was allowed to serve his sentence in the open. He died in Alice Springs.

NATIONAL COUNCIL OF WOMEN OF AUSTRALIA (NCWA) An independent, non-sectarian, non-political organization. The International Council of Women, to which the NCWA belongs, was founded in the United States of America in 1888. The *National Council of Women of New South Wales was formed in 1896. National Councils of Women were later formed in the other Australian states and territories, and the national organization was formed in 1931. Members of the National Councils of Women have worked for causes such as *equal pay, the introduction of policewomen, and uniform divorce laws. Notable members have included Rose *Scott, Vida *Goldstein, Alice *Henry, and Jessie *Street.

NATIONAL COUNTRY PARTY *National Party

NATIONALIST PARTY An Australian political party formed during the First World War. The *National Labor Party and the *Liberal Party merged to form it in early 1917, following the split in the *Australian Labor Party. Its full title was the Australian National Federation, but at times it was also called the National Party. The Nationalist Party, stressing a 'Win-the-War' policy, defeated Labor in the federal elections of 1917. It continued in government until 1929, governing in coalition with the Country Party from 1923 onwards. *Hughes led the party until 1923, when he was forced to resign as leader before the Country Party would join the Nationalists in a coalition. *Bruce succeeded him. The Bruce–*Page Government, many of whose policies were summed up in the slogan *'Men, Money, and Markets', was defeated by Labor in the elections of 1929, and Bruce lost his seat. *Latham succeeded Bruce as leader. The Nationalist Party was absorbed into the newly formed *United Australia Party in 1931. Nationalists also governed in each of the states, either alone or in coalition with other parties, at various times.

NATIONAL LABOR PARTY An Australian political party. During the First World War the *Australian Labor Party split, basically over the issue of *conscription. After the first conscription referendum, *Hughes, the Prime Minister, either left or was removed from (a debatable point) the Labor Party on 14 November 1916. He took with him thirteen Labor Members of the House of Representatives and eleven Labor Senators. Hughes and his followers called themselves the National Labor Party. This rump party governed federally with the support of the *Liberal Party. National Labor and the Liberals merged to form the *Nationalist Party on 9 January 1917.

NATIVE POLICE

First World War cartoon by David Low entitled Patching the Drum, *showing Billy Hughes applying the patch. His **Nationalist Party** won the 1917 federal elections with a 'Win-the-War' policy. (La Trobe Collection, State Library of Victoria)*

NATIONAL PARTY (1917–31) *Nationalist Party

NATIONAL PARTY OF AUSTRALIA An Australian political party that largely represents rural interests. Farmers' representatives were elected to colonial (later state) parliaments from the 1890s onwards, but the First World War acted as a catalyst in the formation of Australian country parties. By the early 1920s they existed (under various names) in all states. A number of farmers' candidates were elected to Commonwealth Parliament in the elections of 1919, and in 1920 they formed the Australian Country Party. It has governed federally in coalition with the *Nationalist Party (1923–9), the *United Australia Party (1934–9, 1940–1) and the *Liberal Party of Australia (1949–72, 1975–83). Its federal parliamentary leaders have included William James McWilliams (1920–1), *Page (1921–39), Archie Galbraith Cameron (1939–40), *Fadden (1941–58), *McEwen (1958–71), John Douglas Anthony (1971–83), Ian McCahon Sinclair (1983–9), Charles W. Blunt (1989–90), and Tim Fischer (1990–). Three have become prime ministers, albeit for very brief periods—Page (1939), Fadden (1941), and McEwen (1967–8). The party has also governed (under various names), mostly in coalition with other parties, for periods in New South Wales, Victoria, Queensland, South Australia, Western Australia and the Northern Territory. The party's name was changed to the National Country Party in 1975, and to the National Party of Australia in 1982.

NATIONAL SERVICE SCHEME *Conscription

NATIVE POLICE (also known as Black police) Forces of Aboriginal police, which were used in some parts of Australia during the nineteenth century. The forces consisted of Black troopers led by White officers. During the period of the *Port Phillip Protectorate, native police were introduced in the Port

Native police forces, members of which are depicted in this Thomas Ham engraving entitled Native Police Encampment, *were established in several Australian colonies during the nineteenth century. (La Trobe Collection, State Library of Victoria)*

Phillip District in 1842, and operated there until 1852. They were re-formed briefly to hunt *Kelly and his gang during the 1870s. Native police were introduced in New South Wales in 1848, and used mainly in the northern districts (part of present-day Queensland). When Queensland became a separate colony in 1859, it inherited the system. Native police continued to operate in Queensland until the late 1890s. The force there usually consisted of between one and two hundred men. Most were used to patrol the frontiers of White settlement. The native police were engaged in many conflicts with other Aborigines, killing many of them.

NAVY *Royal Australian Navy

NCWA *National Council of Women of Australia

NEILSON, JOHN SHAW (1872–1942), poet, was born in Penola, South Australia, the son of a Scottish-born father, who also wrote poetry, and an Australian mother of Scottish descent. He was educated briefly at Penola and Minimay state schools, but spent most of his childhood working as a farm labourer. His family struggled to survive as *selectors in various parts of north-western Victoria from 1891 onwards, supplementing their income by undertaking other work. John Shaw Neilson, whose health was poor, reputedly had more than two hundred labouring jobs in thirty years, many of them in that area. He began to publish poems in newspapers from 1890 and the *Bulletin from 1896. The latter's literary editor, A. G. Stephens, went on to act as Neilson's mentor and agent for many years. After Neilson's eyesight began to fail in about 1905, he was forced to dictate most of his poems to others. Neilson, who received a small Commonwealth Literary Fund pension from 1922 onwards (but not enough to live on), finally worked as an office messenger at the Country Roads Board in Melbourne from 1928 to 1941, when he left on sick leave. After several months in Queensland, he returned to Melbourne, where he later died. The beauty of his work, which includes some of Australia's finest lyrical poetry, stands in stark contrast to the hardship of his life. 'The Orange Tree', inspired by fruit-picking experiences at Merbein in 1917 and published in his *Ballad and Lyrical Poems* (1923), is considered to be his most magical poem. Neilson's other publications include *Heart of Spring* (1923), *New Poems* (1927), *Collected Poems* (1934), and *Beauty Imposes* (1938). Various composers, including Margaret *Sutherland, have set some of his poems to music.

NEW AUSTRALIA was the name given to the first of two Utopian settlements founded by Australian socialists in Paraguay. After the failure of the *Maritime and *Shearers' Strikes of 1890 and 1891, a New Australia Co-operative Settlement Association was formed. About two hundred settlers, led by William *Lane, left Australia on the *Royal Tar* on 16 July 1893. They established New Australia on land provided by the Paraguayan Government at Puesto De Las Ovejas, 175 kilometres south of Asuncion, on 28 September 1893. Another group of settlers arrived on 31 December 1893. In May 1894 Lane, and sixty-three others, left the settlement and founded another in July, called Cosme. An imbalance in the numbers of men and women, together with economic, physical and leadership difficulties, contributed to New Australia's demise. A new agreement with the Paraguayan Government was reached, and the settlement's assets were divided in 1897. Cosme faced similar problems, and its assets were divided in 1929. Descendants of the settlers were still living in Paraguay in the late twentieth century.

NEW AUSTRALIANS A term used in Australia during the late 1940s, the 1950s, and the 1960s to describe migrants, especially those from European countries. *Calwell, Minister for Information and Immigration in the

*Settlers of Cosme, one of two Australian settlements established in Paraguay during the 1890s, the other being **New Australia**. Both proved to be more Utopian in theory than in practice. The tallest woman in the group on the left is Mary Gilmore, later to become famous as a writer. (National Library of Australia)*

*Chifley Labor Government, introduced the term during the period of mass *immigration following the Second World War. Its purpose was to assist migrants to assimilate, and its public use was encouraged. It was designed to replace derogatory terms such as 'Balt', 'Dago', 'Pommy', and 'Reffo'.

NEWCASTLE, a city and port in New South Wales, is situated at the mouth of the Hunter River. John Shortland found the site in 1791 while searching for escaped convicts, named the river, and discovered coal, some of which was exported in 1799. In 1801 small numbers of soldiers and convicts settled at Coal Harbour (as it was then called) to mine coal, but were withdrawn the following year. A *penal settlement existed at the site, renamed Newcastle, from 1804 to 1824, convicts being employed in timber-getting, coal-mining, and lime-burning. The *Australian Agricultural Company became involved in coal-mining at Newcastle in the 1820s, developing a monopoly that lasted until 1850. Newcastle became a municipality in 1859, a city in 1885, and was connected by rail to Sydney (a distance of 168 kilometres) in 1889. The *Broken Hill Proprietary Company Limited opened iron and steel works at Newcastle in 1915, which, along with coal-mining, are still important.

NEW ENGLAND NEW STATE MOVEMENT The most significant of a number of movements towards the formation of new Australian states that have occurred since federation in 1901. The area concerned is that of New England in New South Wales. Earle *Page led the first attempt at Grafton in 1915. The Cohen Royal Commission, appointed by the New South Wales Government, recommended against the formation of any new states in 1925. The New England movement revived during the *Great Depression, it adopted a draft state constitution at Maitland in 1931, and unsuccessfully asked the Commonwealth Government to hold a referendum on the subject. In 1935 the Nicholas Royal Commission, also appointed by the New South Wales Government, reported that New South Wales could be divided into three areas, each of which could vote on the matter at a referendum. The New South Wales Government did not act on this report. The New England movement revived again at Armidale in 1948; Nicholas's suggested boundaries were accepted, and a draft state constitution was adopted in 1949. A constituent assembly of New England was established in 1954, holding its first meeting the following year. A referendum was finally held in the area on 29 April 1967. The proposal was defeated: 198 812 people voted against it, 168 103 in favour of it.

NEW GUARD A quasi-fascist organization formed during the *Great Depression. A small group, led by Eric Campbell, left the Movement, or Old Guard (as it became known), and formed the New Guard at a meeting held at the Imperial Service Club in Sydney in 1931. Several weeks later Campbell was elected leader. At its peak in 1932 the New Guard probably had more than fifty thousand members, many of whom were ex-servicemen. It published two journals, the *New Guard* (1931–2) and the *Liberty* (1932–4). Campbell heavily criticized Jack *Lang, the New South Wales Premier, at a Lane Cove rally in 1932; he was fined £2 for using insulting language, but successfully appealed. In 1932 the New Guard petitioned Philip Woolcott Game, the New South Wales

Governor, for the dismissal of the Lang Government. Francis de Groot, a New Guard member, slashed the ribbon at the opening of the *Sydney Harbour Bridge in 1932; he was fined £5, with £4 costs, for offensive behaviour. Eight New Guard members attacked 'Jock' Garden, one of the founders of the *Communist Party of Australia, in 1932; each was convicted of assault and was sentenced to three months' jail. Police were said to be preparing charges of seditious conspiracy against New Guard leaders shortly before the dismissal of the Lang Government; charges were not pressed. A royal commission and a commission of inquiry were proposed, but neither eventuated. The Centre Party, a political wing of the New Guard, fielded several candidates, including Campbell, at the New South Wales elections of 1935. All were unsuccessful. The New Guard was virtually dissolved by the end of 1935.

NEW GUINEA is a large island in the Pacific Ocean, to the north of Australia. It is divided from Australia by Torres Strait, which was discovered by Luis Vaez de Torres, a Spanish navigator, in 1606. Portuguese navigators were the first Europeans to visit it, during the sixteenth century. It was already inhabited. Spanish, Dutch, British, and French navigators later followed the Portuguese. Permanent White settlement did not begin in New Guinea until the 1870s. New Guinea is now divided, at longitude 141°E, into *Irian Jaya (in the west) and *Papua New Guinea (in the east).

NEW HOLLAND The early name by which the Australian continent was known. Seventeenth-century Dutch navigators named the western part of the continent 'Nova Hollandia', and the name came to be used to describe the entire continent. It was used until about the 1820s, by which time the name 'Australia' was in wide use.

NEW ITALY Popular name given to an Italian settlement near Woodburn on the Richmond River, New South Wales. A group of about two hundred Italian people arrived in Sydney in 1881, after unsuccessfully attempting to settle in New Ireland. Some time later they founded the settlement that became known as New Italy. In 1885 it consisted of fifty-three farms. The settlement lasted into the twentieth century, although some of its members had begun to leave about ten years after its foundation.

NEW NORCIA A Benedictine mission settlement in Western Australia, on the Moore River, about 130 kilometres north of Perth. In 1846 two Spanish Benedictine monks, Dom Salvado and Dom Serra, founded a mission, first known as Central, in the area. In 1847 they chose the present site and name (after St Benedict's birthplace, Norcia, in Italy). Their main aim was to teach Aborigines to become Christians. A church, a monastery, schools, and a farm were established. Aborigines were taught farming and other skills there. The mission was separated from the Diocese of Perth in 1859, and the monastery was made an abbey in 1867. New Norcia continues to exist in the 1990s.

NEW PROTECTION The name given to a particular socio-economic policy. In 1907 *Deakin, one of its chief exponents, described New Protection in the following way:

The 'Old' Protection contented itself

*The church, and some of its members, at the Benedictine mission settlement of **New Norcia** in the 1890s. (J. S. Battye Library of West Australian History)*

with making good wages possible. The 'New' Protection seeks to make them actual. It aims at according the manufacturer that degree of exemption from unfair outside competition which will enable him to pay fair and reasonable wages without impairing the maintenance and extension of his industry, or its capacity to supply the local market. It does not stop here. Having put the manufacturer in a position to pay good wages, it goes on to assure the public that he does pay them.

Such ideas had been discussed in Australia since the 1890s. Deakin's Protectionist Government, supported by the Labor Party, attempted to implement them between 1905 and 1908. It introduced legislation such as the *Excise Tariff (Agricultural Machinery) Act* 1906 and the *Australian Industries Preservation Act* 1906. Such legislation was designed to introduce New Protection by means of bounties, rebates, regulations, and restrictions. The **Harvester Judgement* was an outcome of New Protection. The Labor Party incorporated New Protection into its party policy in 1908. The *High Court found much of the New Protection legislation to be constitutionally invalid, for example, in *Barger's Case* (1908).

NEW SOUTH WALES A state of the *Commonwealth of Australia since the latter's creation in 1901. In 1770 James *Cook took possession for Britain of the eastern part of present-day Australia. In 1786 Britain decided to establish a penal colony, Australia's first White settlement, at *Botany Bay, and in 1788 the *First Fleet arrived at Port Jackson. *Transportation to New South Wales ended in 1840. Partial representative government was granted in 1842; *responsible government was proclaimed in 1855. Originally New South Wales comprised much of Australia. *Van Diemen's Land (present-day Tasmania) was separated from New South Wales in 1825, the

*Port Phillip District (present-day Victoria) in 1851, the *Moreton Bay District (present-day Queensland) in 1859; the Northern Territory was annexed to South Australia in 1863, and land for the *Australian Capital Territory was transferred to the Commonwealth in 1911. *New Zealand was administered from New South Wales in 1840 and 1841.

NEW SOUTH WALES CORPS A unit of the British Army, raised in England in 1789 by *Grose, specifically to police the colony of *New South Wales. It replaced the marines who had sailed with the *First Fleet. The first detachment, consisting of one hundred men commanded by Nicholas Nepean, arrived in New South Wales in 1790, having sailed with the *Second Fleet. John *Macarthur was one of the officers. The second detachment, commanded by Grose, arrived in 1792. Some remaining marines formed a company of the Corps, commanded by George Johnston. The Corps, which has since been dubbed the 'Rum Corps' because of its involvement in the rum trade, greatly increased its power and influence during the interregnum (1792–5). It quelled the *Castle Hill Uprising of 1804. Johnston and the Corps played a major role in the *Rum Rebellion of 1808. As a result, the Corps was replaced by the 73rd Regiment, which some of the men joined; most of the officers, and some of the men, returned to England in 1810. The Corps became the 102nd Regiment when it was recalled to England. Johnston was cashiered in 1811.

*Early White settlers contending with floods on the Hawkesbury River, **New South Wales**. (National Library of Australia)*

NEW UNIONISM is the term used to describe the expansion of the trade union movement in Australia, and changes in its nature, from the late 1870s and 1880s onwards. Before then most trade unions had been small craft unions, restricted to skilled and semi-skilled workers. They charged high membership fees, were local in organization, and were concerned with issues such as hours, pay, and benefits (for example, for sickness). The so-called 'new' unions aimed at mass membership of unskilled and semi-skilled workers in large industries such as shearing and mining. They charged low fees, were inter-colonial in organization, and were more prepared to resort to strikes and political action to gain their demands than the earlier unions had been. The Amalgamated Miners' Association (AMA) and the Amalgamated Shearers' Union (ASU) are examples of 'new' unions formed during the 1880s.

NEW ZEALAND, a country situated in the southern Pacific Ocean, comprises the North Island, the South Island and a number of smaller islands. Its first inhabitants were Polynesians. The first Europeans known to have visited present-day New Zealand were Abel *Tasman (1642), James *Cook (1769–70, 1773, 1777), and the French explorers Jean de Surville (1769), and Marion du Fresne (1772). Europeans seeking seals, whales, timber, and flax went there from the 1790s. Missionaries, notably Samuel *Marsden, began to go there from 1814. There were probably about two thousand European settlers in New Zealand by the late 1830s. A British Resident, James Busby, arrived in 1833. The New Zealand Company (originally Association), which was formed in Britain by Edward Gibbon *Wakefield and others in 1839, established settlements in New Zealand from the 1840s. New Zealand was briefly a dependency of *New South Wales in 1840 and 1841. William Hobson, the first lieutenant-governor of New Zealand, arrived in January 1840, signed the Treaty of Waitangi with Maoris in the North Island in February 1840, and formally annexed New Zealand for Britain in May 1840. New Zealand was proclaimed a separate British colony in 1841, with Hobson becoming its first governor. A form of self-government was granted in 1852, responsible government in 1856, and dominion status in 1907. New Zealand showed some interest in federation with the Australian colonies, but finally did not join. New Zealand ratified the *Statute of *Westminster 1931* in 1947, and is a member of both the *Commonwealth of Nations and the *United Nations Organization. Some Australians took part in the *New Zealand Wars (formerly known as the Maori Wars) during the nineteenth century. Australians and New Zealanders, known as *Anzacs, have fought together at Gallipoli and elsewhere this century. New Zealand signed the *ANZUS Pact with Australia and the USA in 1951, but effectively withdrew from it following its 1984 decision regarding nuclear ships.

NEW ZEALAND WARS (formerly known as the Maori Wars) A series of conflicts in 1860–72 between British troops and Maoris. The basic cause was the struggle for possession of land in the North Island of New Zealand. About 1500 male volunteers sailed from Australia in 1863 to support Britain. They became known as the Waikato Militia (named after the area in which they fought), which consisted of four regiments. A further 1200 or so male volunteers and approximately 1000 women and children followed in 1864. A scheme of military settlement, in which some of the Australians participated, was largely unsuccessful; many of the settlers left their farms.

NINETEEN COUNTIES The official area of White settlement in *New South Wales during the 1830s. The counties, which were proclaimed in the *Sydney Gazette* on 17 October 1829, were Argyle, Bathurst, Bligh, Brisbane, Camden, Cook, Cumberland, Durham, Georgiana, Gloucester, Hunter, King, Murray, Northumberland, Phillip, Roxburgh, St Vincent, Wellington, and Westmoreland. *Squatters (some with authorization, others without) settled beyond the Nineteen Counties.

NOLAN, SIR SIDNEY ROBERT (1917–), artist, was born in Melbourne. His father was a tram-driver. Nolan was educated at state and technical schools in Melbourne, and worked in various jobs before taking up painting in the 1930s. During the Second World War he served in the *army from 1942 to 1945. He was stationed in the Victorian Wimmera, which inspired some of his paintings. Nolan has travelled, lived, worked and exhibited his work in numerous countries. The art critic Alan McCulloch has said of him: 'By 1962 Nolan had achieved a more extensive international reputation than any other Australian painter of his own or preceding generation.' He is perhaps best known for his paintings based on the theme of Ned *Kelly. Many of his other paintings are of Australian historical and outback subjects. Nolan has been awarded numerous prizes and fellowships, including the Dunlop Australian Art Contest (1950), an Italian Government scholarship (1956), a Commonwealth Fund Fellowship to the USA (1958), and the Britannica Australian Award (1969).

NORFOLK ISLAND, a small Pacific Island, has been an Australian external territory since 1914. James *Cook was the first White person to discover the island, in 1774. It was uninhabited at the time. *Phillip established a convict settlement there, as an outpost of New South Wales, in 1788. Attempts to develop a flax industry were unsuccessful, and the first settlement was abandoned in 1814.

*A chain gang on **Norfolk Island**. (National Library of Australia)*

A second, mainly for convicts undergoing secondary punishment, was established in 1825. Alexander *Maconochie experimented in penal reform there from 1840 to 1844. The second settlement was finally abandoned in 1856, the year in which Pitcairn Islanders (descendants of *Bounty* mutineers and Tahitian women) settled on the island. Many of their descendants still live there.

NORTH AUSTRALIA (also known as the Gladstone Colony) was a short-lived British colony on the central coast of present-day Queensland. The British Secretary of State for the Colonies, William Ewart Gladstone, established it in 1846. The colony was to comprise all of New South Wales (as it then was) north of latitude 26°S. Its population was to consist of expirees (convicts whose terms of transportation had expired) and conditionally pardoned convicts from Van Diemen's Land, and *exiles from Britain. There was vigorous opposition in New South Wales to the proposal. The new colony's Superintendent, George Barney, chose Port Curtis as the site for the capital in November 1846. He and the other first settlers, who included officials, soldiers, and civilians, arrived there from Port Jackson in January 1847. More soldiers and civilians arrived soon after. The settlers at Port Curtis faced a number of problems, mostly relating to the climate. Meanwhile, Earl Grey, who had succeeded Gladstone as Secretary of State for the Colonies, wrote to Governor FitzRoy in November 1846, telling him of the British Government's decision to abandon the new settlement. This news reached Port Curtis in April 1847, and the settlement was abandoned shortly after. The town (later city) of Gladstone was later established on the site.

NORTHERN TERRITORY A self-governing territory of Australia. The Dutch ship *Arnhem*, the first known European contact, visited the coast in 1623. Abel *Tasman (1644), Nicholas *Baudin (1801, 1803), Matthew *Flinders (1803), and Phillip Parker King (1818) followed. Several attempts were made to establish a permanent British commercial and strategic settlement in the north of Australia, partly because of fears that the French or others might settle in the area. The first party from New South Wales arrived at *Port Essington in 1824, took formal possession of the area, and founded Fort Dundas on Melville Island. Fort Dundas was abandoned in 1829. Fort Wellington, at Raffles Bay, lasted from 1838 to 1849. *Stuart crossed Australia from Adelaide to the northern coast in 1862, and the Northern Territory was annexed to South Australia in 1863. Palmerston was founded at Escape Cliffs in 1864, but this site was abandoned in 1867. A new site was chosen for Palmerston in 1869, and this settlement, later renamed *Darwin, proved to be permanent. The Commonwealth took over the administration of the Northern Territory in 1911. The Territory was divided into North Australia and Central Australia, each separately administered, between 1926 and 1931. A legislative assembly was created in 1974, and the Northern Territory was granted self-government in 1978.

NOVAR, RONALD CRAUFURD MUNRO-FERGUSON, VISCOUNT (1860–1934), governor-general, was born in Scotland, and privately educated. He served in the Grenadier Guards from 1879 to 1884, was a Member of Parliament for many years (1884–5, 1886–1914), and was made a Privy Counsellor in 1910. As Sir Ronald Craufurd Munro-

Ferguson, he was Governor-General of Australia from 1914 to 1920. He was created a viscount in 1920, and was the Secretary for Scotland from 1922 to 1924.

NUCLEAR TESTS were conducted in Australia by the British Government, with the agreement of the *Menzies Government, between 1952 and 1963. Three bombs were exploded in the Monte Bello Islands in 1952 and 1953, two at Emu in 1953, and seven at Maralinga in 1956 and 1957. (The latter two sites are near *Woomera.) Other minor related trials took place between 1953 and 1963. In 1984 and 1985, a royal commission headed by Mr Justice James McClelland inquired into the tests. Its report, published in December 1985, criticized Menzies, found that the tests had not been safely conducted, that some Aborigines had been subjected to radiation, and that some lands were still contaminated, and recommended that the Australian Government pay compensation to the Aborigines involved and the British Government pay for the removal of the contamination.

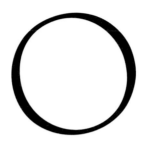

OLYMPIC GAMES The first of the modern Olympic Games, international sporting contests usually conducted every four years, was held in Athens in 1896. Australians have participated in each of the modern Olympic Games, and Melbourne hosted those in 1956, in which Australia was placed third, having won thirteen gold, eight silver and fourteen bronze medals, behind Russia, and the USA.

O'MALLEY, KING (1858?–1953), insurance salesman and politician, claimed to have been born in Quebec, Canada, but was probably born in Kansas, USA, the son of an Irish rancher and soldier and his wife. He was educated in New York, and worked for some time in his uncle's bank there, before working as an insurance and real-estate agent in various American states from about 1880 onwards. O'Malley, a flamboyant and controversial character, allegedly created his own church (for financial reasons), definitely became a teetotaller (opposed to 'stagger juice'), and reputedly worked as a journalist on the *Arizona Kicker*. After arriving in Melbourne in 1888, he sold insurance in Victoria, Tasmania, Western Australia, and South Australia for a number of years. Despite some initial problems regarding his nationality (only British citizens being eligible to stand), he was an Independent Member of the South Australian House of Assembly from 1896 to 1899. O'Malley, a supporter of the *federation movement, was an Independent, then a Labor Member of the House of Representatives, representing Tasmanian electorates from 1901 to 1917. He was a member of the Select Committee (1904), and then the Royal Commission (1905–6), on Old Age Pensions. He was Minister for Home Affairs from 1910 to 1913 and again from 1915 to 1916. O'Malley is considered to have been largely responsible for the establishment of the Commonwealth Bank (later the *Commonwealth Banking Corporation) and played an important role in the planning of *Canberra. A pacifist, O'Malley was defeated in the 1917 federal elections, held some months after the *Australian Labor Party split over the issue of *conscription. He stood unsuccessfully as an Independent Labor candidate for the House of Representatives in 1919 and 1922. He spent the rest of his life in Melbourne, concentrating on his investments. Supporters of women's rights, O'Malley and his second wife left money for girls' scholarships.

ORD RIVER IRRIGATION SCHEME A project to use the Ord River (named after Sir Harry Ord, the Governor of Western Australia from 1878 to 1880) to irrigate a large area in the previously largely uninhabited Kimberley region in the north of Western Australia. Work began on the scheme in 1960, a township named Kununurra was established in 1963, and Australia's largest artificial lake, the Ord River Dam, enclosing Lake Argyle, was opened in 1972. The lake covers the Argyle pastoral property, which was previously owned by the well-known Durack family. Some critics claimed that the scheme, which was largely funded by the Commonwealth Government, was politically motivated. Attempts have been made to grow various crops, including cotton, rice, and sugar, but the scheme is generally considered to have been an expensive disappointment. Agricultural and economic problems have been blamed. Diamonds were found in the area in 1979.

*One of the best-known pictorial representations of **overlanders** is S. T. Gill's water-colour of that name (c. 1865). (Art Gallery of New South Wales)*

ORR CASE An Australian academic *cause célèbre* of the 1950s and 1960s, involving Sydney Sparkes Orr (1914–66), who had been appointed professor of philosophy at the University of Tasmania in 1952. Years of discontent at the University over issues such as salaries and conditions culminated in the publication in the Hobart *Mercury* on 29 October 1954 of an open letter to the Tasmanian Premier written by Professor Orr and thirty-five other staff members, calling for an inquiry. A Tasmanian royal commission was then held. Its June 1955 report criticized all parties involved and recommended that the University Council be reconstituted, but little was done about it. On 16 March 1956 the University dismissed Professor Orr as the result of several charges, the most serious of which involved sexual misconduct with one of his students, Suzanne Kemp. Orr sued for wrongful dismissal, but in October 1956 the Tasmanian Supreme Court found him guilty of the sexual misconduct charge (but not the others) and ruled that the dismissal was justified. His subsequent appeal to the *High Court was dismissed in May 1957. Australian academics were divided over the case. In August 1958 the Australasian Association of Philosophers placed a black ban on Orr's chair and it was not filled again until 1969. The Federation of Australian University Staff Associations conducted its own inquiry, decided in February 1961 that the University had denied Orr natural justice and that the Supreme Court judgement had been wrong, and censured the University. In February 1966 the University paid Orr a settlement of £16 000 plus some costs, but did not re-employ him. Orr, who had been in poor health, died in July 1966.

'OUR LAST MAN AND OUR LAST SHILLING' A phrase used before, but made famous during, the First World War. Andrew *Fisher promised that Australia would defend Britain to 'our last man and our last shilling' in several speeches given during the federal election campaign that coincided with the outbreak of war. As newly elected Prime Minister, he promised: 'We shall pledge our last man and our last shilling to see this war brought to a successful issue.'

OVERLANDERS Popular name given to Australian stockmen, especially during the nineteenth century, who drove cattle and sheep over very long distances. The journeys of the early overlanders into areas of Australia previously unoccupied by White settlers, usually took months, sometimes years. Overlanding from New South Wales to the *Port Phillip District began in 1836; from New South Wales and the Port Phillip District to South Australia in the late 1830s; from New South Wales to the *Moreton Bay area in 1840; and across northern Australia in the 1880s. The main routes used in the late nineteenth century were the Murranji Track, from the Northern Territory to Queensland; the Birdsville Track, from Queensland to South Australia; and a track from Queensland to Victoria. Droving along these and other routes continues, but has declined since the development of rail and road transport, and refrigeration.

OVERLAND TELEGRAPH LINE Australia's first trans-continental telegraph system, completed in 1872. Charles (later Sir Charles) Todd, the South Australian Postmaster-General and Superintendent of Telegraphs, was largely responsible for the construction of the line, running from Port Augusta in South Australia to Port Darwin in the Northern Territory, a distance of about 2900 kilometres. The South Australian Government paid for the line, which enabled Australia for the first time to be linked (via Java) with the telegraph systems of Asia and Europe. The Overland Telegraph Line played an important role in Australia's overseas communication system for about ninety years.

*The construction of the **Overland Telegraph Line**, stretching from Port Augusta in the south to Darwin in the north, helped to break down Australia's isolation from the rest of the world. (La Trobe Collection, State Library of Victoria)*

P

PACIFIC ISLANDERS *Blackbirding

PAGE, SIR EARLE CHRISTMAS GRAFTON (1880–1961), politician, was born at Grafton, New South Wales, educated at Grafton State School, Sydney High School, and the University of Sydney, and then practised as a surgeon. During the First World War he served in the Australian Army Medical Corps in Egypt and France from 1916 to 1918. Early in the war Page had been active in the *New England new state movement; after the war he entered federal politics. He was a Member of the House of Representatives from 1919 to 1961, representing the Country Party (later the *National Party of Australia) from 1920, and leading it from 1921 to 1939. Page was Deputy Prime Minister in coalition governments from 1923 to 1929, and from 1934 to 1939, and briefly Prime Minister following the death of *Lyons in 1939. He was Treasurer from 1923 to 1929, and was responsible for the establishment of the Loan Council, which he chaired from 1924 to 1929, and for the introduction of the *Financial Agreement. He was Chancellor of the University of New England from 1955 to 1960. Page was defeated at the 1961 elections. He died during the election count.

PALMER, EDWARD VIVIAN (VANCE) (1885–1959), writer and critic, was born in Queensland. His father was a schoolteacher. Palmer was educated at Ipswich Grammar School; he then worked in various jobs, including those of clerk, journalist, tutor, and book-keeper, in Australia and overseas. During the First World War he enlisted with the AIF, but did not see action. He spent most of his life in Australia. Like his wife, Janet (Nettie) *Palmer, he did much to foster Australian literature. He wrote poetry, prose and drama; he helped to found a dramatic group, the Pioneer Players, which existed from 1922 to 1926; he was active on the Advisory Board of the Commonwealth Literary Fund, which he chaired from 1947 to 1953; and he encouraged other writers. His poetry includes *The Forerunners* (1915), his collections of short stories *Sea and Spinifex* (1934), his novels the trilogy made up of *Golconda* (1948), *Seedtime* (1957), and *The Big Fellow* (1959), his plays *The Black Horse and Other Plays* (1924) and *Hail Tomorrow* (1947), and his critical works *National Portraits* (1940), and *The Legend of the Nineties* (1954).

PALMER (née Higgins), JANET (NETTIE) GERTRUDE (1885–1964), writer and critic, was born in Victoria. She was educated at the Presbyterian Ladies College, Melbourne, and at the University of Melbourne. She married Edward Vivian (Vance) *Palmer in 1914, and spent most of her life in Australia. Nettie Palmer actively encouraged the development of Australian literature. She wrote in various genres, lectured for organizations such as the Commonwealth Literary Fund and the Melbourne University Extension Board, and encouraged other writers. She died in Melbourne. Her poetry includes *The South Wind* (1914) and *Shadowy Paths* (1915). Her critical works include *Modern Australian Literature, 1900–23* (1924), *Henry Bournes Higgins* (1931), *Talking It Over* (1932) and *Henry Handel Richardson* (1950). Her *Fourteen Years* (1948) is autobiographical. She also edited a collection of twentieth-century Australian short stories entitled *Australian Storybook* (1928).

*Earle Christmas Grafton **Page** was the leader of the federal parliamentary Country Party (later the National Party of Australia) from 1921 to 1939. (National Library of Australia)*

PALMERSTON

Phil May's cartoon The Mother of Civilization *(1888), based upon a comment made by Henry Parkes. (La Trobe Collection, State Library of Victoria)*

'New South Wales is the mother of civilization in this part of the world.' Phil May (Parkes' Speech)

PALMERSTON *Darwin

PAPUA NEW GUINEA An independent member of the *Commonwealth of Nations. It comprises the eastern part of the island of *New Guinea and nearby islands. Britain rejected Queensland's annexation in 1883 of south-eastern New Guinea, proclaimed a protectorate over this area in 1884, and finally annexed it in 1888, after Queensland, New South Wales, and Victoria had agreed to contribute financially to its administration (which was to be handled by Britain and Queensland). British New Guinea was transferred to the newly created Commonwealth of Australia in 1902, which formally took control in 1906; British New Guinea became Papua. Germany had administered north-eastern New

Guinea from 1884 onwards, but Australia captured it in 1914, at the beginning of the First World War, and administered the Territory of New Guinea, as it became known, under a *League of Nations mandate from 1921 and under United Nations trusteeship from 1946. Japan occupied much of New Guinea during the Second World War. Australia administered Papua and the Territory of New Guinea together from 1946 onwards. The two territories were formally united as Papua New Guinea in 1949. Papua New Guinea gained self-government in 1973 and independence in 1975.

PARIS PEACE CONFERENCE A meeting of the Allied and Associated Powers held in Paris in 1919–20, shortly after the end of the First World War, to determine the peace settlement. The provisions of the Treaties of *Versailles, St Germain, Neuilly, Trianon, and Sevres were determined. In a significant move for Australia, the dominions were represented at the Conference. Billy *Hughes, as one of Australia's two representatives, played an active and controversial role in the proceedings.

PARKES, SIR HENRY (1815–96), politician, was born in England. His father was a tenant farmer, later a gardener. Parkes's formal education was limited. He held labouring jobs before serving an apprenticeship as a bone and ivory turner, then founded a business, which, like several of his later businesses, failed. Parkes and his wife sailed to Sydney as bounty immigrants in 1839. He worked as a labourer, then in an ironmongery and brass foundry, and then in the Customs Department from 1840 to 1844. During the 1840s Parkes contributed to a number of newspapers, including the *Atlas*, the *Australasian Chronicle*, the *Launceston Examiner* and the *Sydney Morning Herald*. He established, edited and owned the *Empire* newspaper, which was published from 1850 to 1858. Parkes's political career was extensive. He was active in a number of political campaigns (notably ones opposing *transportation, and supporting both Robert Lowe and John Dunmore *Lang), before serving as a Member of the Legislative Council of New South Wales (1854–6). He was then a Member of the colony's Legislative Assembly (1856, 1858, 1859–61, 1864–70, 1872–84, 1885–95). He was Colonial Secretary (1866–8, 1872–5, 1877, 1878–83, 1887–89, 1889–91), Premier (1872–5, 1877, 1878–83, 1887–89, 1889–91), and Secretary for Lands (1883). Parkes lectured on *immigration in England between 1861 and 1863. His role in the *federation movement earned him the title 'Father of Federation', but there has been some speculation about his motives. He eschewed the Federal Council, but later declared his support for federation, notably in his *Tenterfield Oration. The Australasian Federation Conference discussed his proposals in 1890. He chaired the *Federal Convention in Sydney in 1891. Parkes died in New South Wales. He wrote poetry and prose, including *Australian Views of England* (1869), *Studies in Rhyme* (1870), *The Beauteous Terrorist and Other Poems* (1885), *The Federal Government of Australasia* (1890), and *Fifty Years in the Making of Australian History* (1892).

PARLIAMENTS The term commonly used to describe the legislatures of each of the states (formerly colonies) and the *Commonwealth of Australia. Legislative councils with nominated members first met in New South Wales in 1823,

Van Diemen's Land in 1825, Western Australia in 1832, and South Australia in 1842. Councils with nominated and elected members first met in New South Wales in 1843, Victoria, South Australia and Van Diemen's Land in 1851, and Western Australia in 1870. Bicameral parliaments (those with two legislative chambers), based partly on the British system, first met in Victoria, New South Wales, South Australia and Tasmania (formerly Van Diemen's Land) in 1856, Queensland in 1859, and Western Australia in 1890. Members of the lower houses (called the Legislative Assembly in New South Wales, Victoria, Queensland and Western Australia, and the House of Assembly in South Australia and Tasmania) were, and are, elected. Male *suffrage for these houses was introduced in South Australia in 1856, Victoria in 1857, New South Wales in 1858, Queensland in 1859, Western Australia in 1893, and Tasmania in 1900. Female suffrage was introduced in South Australia in 1894, Western Australia in 1899, New South Wales in 1901, Tasmania in 1903, Queensland in 1905 and Victoria in 1908. Members of the Western Australian Legislative Council were nominated between 1890 and 1893, and then elected. Members of the New South Wales Legislative Council were nominated until 1934, then indirectly elected until 1978, then directly elected. Members of the Queensland Legislative Council, which was abolished in 1922, were nominated. Members of the legislative councils in Victoria, South Australia, Western Australia (from 1893), and Tasmania, were elected, but a special franchise system (based on qualifications such as property, profession, education, or age) operated for many years, eventually being abolished in all four states between 1950 and 1973. According to the Constitution, the federal parliament consists of 'the Queen, a Senate, and a House of Representatives'. The Senate and the House of Representatives first met in 1901. Members of these two houses have always been elected. Those people eligible to vote for the lower house in each state were eligible to vote for both houses of the federal parliament at the first federal election, full adult suffrage for both houses being introduced in 1902. Methods of dealing with deadlocks between the two houses vary in some of the parliaments.

PARRAMATTA, now a city that forms part of the Sydney metropolitan area, is situated about 24 kilometres west of Sydney Cove. Governor *Phillip, who explored the area in April 1788, established a farming settlement there, the second White settlement in Australia, in November 1788. The settlement was originally known as Rose Hill; its name was changed to Parramatta (an Aboriginal name that probably means 'head of the river' or 'head of the waters') in 1791.

PATERSON, ANDREW BARTON ('Banjo') (1864–1941), poet, was born near Orange, New South Wales. His father was a grazier. Paterson was educated at Sydney Grammar School and the University of Sydney, practised as a solicitor for some time, and then became a journalist. He was a war correspondent for the *Sydney Morning Herald* during the *Boer War and the *Boxer Rising. After his return to Sydney, Paterson continued in journalism; he edited the Sydney *Evening News* from 1904 to 1906, and the Sydney *Town and Country Journal* from 1907 to 1908. He was then a pastoralist for about six years. During the First World War he served as an ambulance officer in France and as a remount officer in Egypt. After the war he again returned to journalism and other writing, sometimes using the pseudonym 'The Banjo', and he is still usually referred to as 'Banjo' Paterson. He is best known for his bush ballads, which attained wide popularity. His work (much of which was first published in the *Bulletin) includes The Man from Snowy River and Other Verses (1895), and Saltbush Bill, J.P., and Other Verses (1917). Paterson wrote the words of the famous Australian song, *'Waltzing Matilda', in 1895. His other publications include the autobiographical Happy Dispatches (1934).

PATERSON, WILLIAM (1755–1810), colonial adminstrator, was born in Scotland, pursued a military career, helped to raise the *New South Wales Corps, and arrived in New South Wales in 1791. Shortly afterwards he went to *Norfolk Island, where he commanded a detachment of the Corps until 1793. Paterson took control of the colony, following Francis *Grose, from December 1794 to September 1795. Ill health forced Paterson to leave New South Wales in 1796, but he returned in 1799, when Philip Gidley *King appointed him Lieutenant-Governor. Paterson administered the new settlement of Port Dalrymple in *Van Diemen's Land from 1804 to 1809. He returned to Sydney in the aftermath of the *Rum Rebellion, took over from Joseph *Foveaux, and remained in control from early in 1809 until Lachlan *Macquarie took office early in 1810. Paterson sailed for England with the New South Wales Corps in 1810, but died while still at sea. He had also been an active botanist.

PATRIOTIC SIX   The popular name given to six non-official members of the *Van Diemen's Land Legislative Council who were involved in political controversy in 1845. They were Richard Dry, Michael Fenton, Thomas Gregson, William Kermode, John Kerr, and Charles Swanston. Van Diemen's Land was still experiencing the *depression of the 1840s. The Six clashed with *Eardley-Wilmot, the Lieutenant-Governor, over annual estimates of expenditure for 1846. They opposed the use of local revenue for the payment of police and jails, believing that Britain should bear the cost, because it was caused by British *transportation. The Six walked out of the Council on 31 October, leaving it without a quorum, and resigned shortly afterwards. Eardley-Wilmot replaced them with other people, but was himself recalled in 1846, allegedly for neglecting the convict system. The Six were later reinstated, but the British Government's policy remained unchanged.

PAYMENT OF MEMBERS   The principle of payment of members of the lower houses of parliament was introduced in Victoria (on a temporary basis for some years) in 1870, Queensland in 1886, South Australia in 1887, New South Wales in 1889, Tasmania in 1890, and Western Australia in 1900. The payment of members of the upper houses followed later. It has always existed for the Commonwealth House of Representatives and Senate.

PENAL SETTLEMENTS   Settlements to which convicts who had committed further crimes within the Australian colonies were sent. *Newcastle in New South Wales was used as a penal settlement from 1804 to 1824; the first convicts sent there were some of those involved in the *Castle Hill Uprising.

*In a dramatic scene during the **Petrov affair**, Mrs Petrov is escorted to an aeroplane at Sydney Airport by Soviet officials. She subsequently defected after leaving the flight in Darwin. (News Limited)*

Port Macquarie (1821–30) and *Moreton Bay (1824–39) were also used as penal settlements. *Norfolk Island was resettled as a penal settlement in 1825; the last convicts left there in 1856. A penal settlement was established at *Macquarie Harbour in Van Diemen's Land in 1822. Building of *Port Arthur began in 1830, and the last convicts from Macquarie Harbour were transferred there several years later. Port Arthur, perhaps the best-known penal settlement, was not closed until 1877.

PERTH, the capital city of *Western Australia, lies on the banks of the Swan River, some 16 kilometres from the mouth of the river. British free settlers, led by *Stirling, founded Perth in 1829. Perth was proclaimed a city in 1856. Its population was 36 274 in 1901, 348 647 in 1954, and 1 118 800 in 1988.

PETROV AFFAIR A controversial diplomatic incident that occurred in Australia in 1954–5, during the Cold War. Vladimir Petrov, of the Soviet Embassy in Canberra, defected on 3 April 1954. He gave documents and information relating to alleged Soviet espionage in Australia to the *Australian Security Intelligence Organization. Shortly after, *Menzies announced news of the defection and that a royal commission into espionage would be held. Petrov's wife, Evdokia, also of the Embassy, defected on 20 April. Diplomatic relations between Australia and Russia ceased, and were not resumed until 1959. The Petrov affair became an issue in the campaign for the federal elections of May 1954, which saw the return of the Menzies Government. The Royal Commission's report was tabled in 1955. It found that the Petrov papers were authentic, and that there had been leaks from the Department of External Affairs between 1945 and 1948, but none after 1948. No prosecutions were launched. *Evatt, who represented two of his staff members before the Royal Commission until his leave to appear was withdrawn, claimed that Menzies had used the defections for electoral purposes, which Menzies denied. The Petrov affair has been seen as a catalyst in the split of the *Australian Labor Party in the mid-1950s.

PHAR LAP (1926–32), Australia's most famous racehorse, was foaled in New Zealand, but raced mostly in Australia. He won many races, including the *Melbourne Cup of 1930. His death in California, shortly after he had won the Agua Caliente in Mexico, was probably due to colic, but it aroused much suspicion in Australia. At the time of his death, Phar Lap was said to be the third highest stake winner in the world, his winnings totalling £66 738. Phar Lap has become part of Australia's folklore.

PHILLIP, ARTHUR (1738–1814), *governor, was born in London. His father was a language teacher. Phillip pursued a naval career; after completing an apprenticeship, he saw active service in the Seven Years' War. During his naval career he served with both the Portuguese and British navies. Phillip was involved in survey work when he was appointed the first governor of New South Wales in 1786; he sailed with the

Phar Lap winning the Melbourne Cup in 1930. (La Trobe Collection, State Library of Victoria)

Pinjarra, Battle of

Captain Arthur Phillip, the first governor of New South Wales, from a painting by F. Wheatly, engraved by W. Sherwin, and published in London in 1789. (Rex Nan Kivell Collection, National Library of Australia)

*First Fleet in 1787. He decided that the landing would be made at Port Jackson, rather than at *Botany Bay as intended, in 1788. Phillip's commissions and instructions gave him virtually autocratic power. He took control of all the continent east of the 135th meridian and the adjacent islands (including *Norfolk Island). His major difficulties in establishing the new settlement included the personalities of some of the settlers (both officers and convicts), the settlers' inadequate farming knowledge and experience, and a lack of equipment and supplies. Phillip sailed for England in 1792 to seek medical treatment, and resigned on medical advice in 1793. He resumed his naval career in 1796, retired in 1805, and spent the rest of his life at Bath. Phillip was promoted to admiral shortly before his death.

PINJARRA, BATTLE OF   A violent conflict on 28 October 1834 between twenty-five Whites (police, soldiers, and settlers, led by Governor *Stirling) and about seventy or eighty Aborigines in the Murray River district (near present-day Pinjarra) of Western Australia. The Whites attacked the Blacks in retaliation for several killings and attacks on property. One White was killed, and another injured; estimates of Black deaths range from about fifteen to fifty.

PLAYFORD, SIR THOMAS (1896–1981), politician, was born at Norton's Summit, South Australia. His grandfather, Thomas Playford (1837–1915), had been the Premier of South Australia twice (1887–9, 1890–2). Playford was a Member of the South Australian House of Assembly from 1933 to 1967, when he retired. He was the Premier of South Australia, leading Liberal and Country League Ministries, from 1938 to 1965, a record continuous term for a parliamentary leader in the *Commonwealth of Nations. (An electoral system favouring rural electorates helped him to stay in office for such a long period.) Playford also held various other portfolios. He encouraged the growth of secondary industry, opened the Leigh Creek coalfield, and nationalized electricity supplies. South Australia's population increased markedly during his term of office. After the Labor Party defeated his Government at the elections of 1965, Playford led the Opposition until 1966.

PLURAL VOTING   A system in which some electors can vote more than once. Plural voting at elections for the lower houses of parliament was abolished in New South Wales in 1893, Victoria in 1899, Tasmania in 1901, Queensland in 1905, and Western Australia in 1907. It had not existed for the lower house in South Australia. It has never existed at elections for the Commonwealth House of Representatives or Senate.

POINT PUER   *Port Arthur

POLICE STRIKE   A group of policemen in Melbourne refused to go on beat duty on 31 October 1923. By 2 November more than six hundred metropolitan policemen were on strike. The total Victorian police force numbered about one thousand eight hundred at the time. The introduction of a system of special supervisors, nicknamed 'spooks' by the policemen, was the immediate cause of the strike, but there was a background of other grievances. Melbourne did not have a proper police force for a week, during which time rioting, looting, and illegal gambling occurred in the streets. A *Public Safety Preservation Act* was passed, giving the State Government

*View of a Melbourne street during the **police strike** of 1923. (National Library of Australia)*

wide emergency powers, and special policemen were recruited. Victorian Government requests to the Commonwealth Government for troops were not granted, except to guard Commonwealth property. Six hundred and thirty-four policemen were discharged, and two dismissed; none was reinstated, but special supervisors were not used again. The Police Strike led to a royal commission chaired by *Monash, and investigation into the police force.

PORT ARTHUR is an inlet, about 100 kilometres by road from Hobart, on Tasman's Peninsula in south-eastern Tasmania. A *penal settlement for male convicts undergoing secondary punishment (that is, those convicts who had committed further crimes within the colonies) was founded there in 1830. One of the site's main advantages was that it was only a short distance by sea from Hobart. Port Arthur continued as a prison after *transportation ended, finally being abandoned in 1877. A boys prison, established at nearby Point Puer in 1834, continued until 1849. At its peak it held about 730 boys. Some of the buildings at Port Arthur have been restored, and the area has become a major tourist attraction.

PORT DALRYMPLE *Launceston

PORT ESSINGTON, an inlet on the Cobourg Peninsula, was the site of a short-lived British settlement, the third White settlement in the area that now forms the *Northern Territory. Like its predecessors, Fort Dundas (1824–9) on Melville Island and Fort Wellington (1827–9) at Raffles Bay, this settlement, which became known as Victoria, was established mainly for commercial and strategic reasons. Captain James John Gordon Bremer had landed at Port Essington in 1824, but had then decided

*Photographs of convicts at **Port Arthur**, taken in about 1870, with brief details of their current sentences. (Queen Victoria Museum and Art Gallery, Launceston)*

James Conolly (or Connolly), serving ten years for attempted unnatural act.

William ('Clocky') Clemo, serving seven years for indecent assault.

Richard Cobbett (or Corbett), serving fourteen years for housebreaking.

Charles Downes, serving life sentence for rape.

Nathan Hunt, serving seven years for larceny.

John Merchant, serving sixteen years for sheep-stealing.

to establish the first settlement on Melville Island instead, because of a lack of fresh water at Port Essington. However, he returned to Port Essington and established a settlement there in 1838. The tropical climate led to many problems for the settlers, and the settlement was abandoned in 1849. A fourth White settlement, at Escape Cliffs (1864–7) near the mouth of the Adelaide River, also failed.

PORT JACKSON *Botany Bay

'PORT OF PEARLS' *Broome

PORT PHILLIP ASSOCIATION An organization formed by fifteen people, including John Batman, in *Van Diemen's Land in 1835. It was originally known as the Geelong and Dutigalla Association, but soon became known as the Port Phillip Association. Batman entered into a so-called treaty with local Aborigines in 1835, exchanging various items, including tomahawks, shirts, and blankets, for some 243 000 hectares of land. Batman and other members of the Association settled in the area, which became known as the *Port Phillip District. The British Government did not accept the validity of the treaty, but did allow the association a remission of £7000 on land purchase price, as recognition of its initial expenses in settling the area. The Port Phillip Association was superseded by the Derwent Association, which lasted until 1842.

PORT PHILLIP DISTRICT The name given to present-day *Victoria in its early years of White settlement. *Bass, *Flinders, James Grant, and John Murray explored its coast in the years between 1797 and 1802. *Collins, sent from England, unsuccessfully attempted to establish a convict settlement on the eastern coast of Port Phillip (near present-day *Sorrento) in 1803, but soon moved to *Van Diemen's Land. The *Hume and Hovell expedition explored parts of the area in 1824. Another convict settlement, on the eastern coast of Western Port, lasted from 1826 to 1828. Three attempts at settlement from Van Diemen's Land proved to be more permanent. *Henty settled at Portland Bay in 1834; John Batman and John Pascoe Fawkner established separate settlements on the site of present-day *Melbourne in 1835; *squatters settled *Australia Felix soon after its discovery by Thomas *Mitchell in 1836. *La Trobe became Superintendent of the Port Phillip District in 1839. A small number of convicts went to the Port Phillip District from New South Wales and Van Diemen's Land. Some of them were employed on public works, others accompanied settlers as assigned servants. A few *exiles, most of whom were employed by squatters, were sent there in the late 1840s. The Port Phillip District was part of New South Wales until 1851, when it became a separate colony, named after Queen Victoria.

PORT PHILLIP PROTECTORATE An organization designed to control and protect Aborigines in the *Port Phillip District. It was the most elaborate of a number of protectorates (others operated in South Australia and Western Australia). The British Government established the Port Phillip Protectorate in 1838, colonial land revenue being used to fund it. A chief protector, *Robinson, and four other protectors were appointed. Each was to cover a different area. They were to travel with the Aborigines, learn their language, protect them and their

Port Phillip Protectorate

*Newspaper headlines announcing the welcome news of the **Port Phillip District**'s separation from New South Wales. (La Trobe Collection, State Library of Victoria)*

property, and persuade them to settle on reserves where they would live like Europeans. The protectors established the reserves, which had farms and schools, and dispensed rations and religious instruction. *Squatters opposed the Protectorate, Aborigines generally did not wish to settle on reserves, and lack of finance was a problem. The Legislative Council of New South Wales abolished the Port Phillip Protectorate in 1849.

PREMIERS' PLAN *Great Depression

PRICHARD, KATHARINE SUSANNAH (1883–1969), writer, was born at Levuka, Fiji, where her father was the editor of the *Fiji Times*. (He returned to Australia in 1886 to edit the Melbourne *Sun*, later the Launceston *Daily Telegraph*, and then the *Australian Mining Standard*.) Prichard was educated at the South Melbourne College, before becoming a governess, and then a journalist, writing for Australian and English newspapers. She travelled widely throughout her lifetime. Her first novel appeared in 1915, and she married Hugo Vivian Hope Throssell in 1919. Katharine Susannah Prichard helped to found the *Communist Party of Australia, was a lifelong member, and wrote several political works, including *Why I am a Communist* (1957). She died at her home near Perth. Her works of fiction, many of which have distinctively Australian settings and are concerned with social justice, include *Working Bullocks* (1926), *Coonardoo* (1929), and *Haxby's Circus* (1930). Her other publications include *Clovelly Verses* (1913), *The Earth Lover and Other Verses* (1932), and an autobiography entitled *Child of the Hurricane* (1963).

PRIME MINISTER The term used to describe the chief minister of the Federal Cabinet. Australia's prime ministers have been Edmund *Barton (1901–3), Alfred *Deakin (1903–4, 1905–8, 1909–10), John Christian *Watson (1904), George Houston *Reid (1904–5), Andrew *Fisher (1908–9, 1910–13, 1914–15), Joseph *Cook (1913–14), William Morris *Hughes (1915–23), Stanley Melbourne *Bruce (1923–9), James Henry *Scullin (1929–32), Joseph Aloysius *Lyons (1932–9), Earle Christmas Grafton *Page (1939), Robert Gordon *Menzies (1939–41, 1949–66), Arthur William *Fadden (1941), John *Curtin (1941–5), Francis Michael *Forde (1945), Joseph Benedict *Chifley (1945–9), Harold Edward *Holt (1966–7), John *McEwen (1967–8), John Grey *Gorton (1968–71), William *McMahon (1971–2), Edward Gough *Whitlam (1972–5), John Malcolm *Fraser (1975–83), Robert James Lee *Hawke (1983–91), and Paul John *Keating (1991–). The chief minister of each state is known as the premier.

PROBATION *Transportation

PROTECTION The 'theory or system of protecting home industries against foreign competition by imposing duties or the like on foreign productions'. Victoria pursued protectionist policies from the 1860s onwards, while New South Wales pursued *free-trade policies. The other colonies adopted various policies. The Commonwealth Government introduced a compromise tariff in 1902, and a national policy or tariff protection in 1908. From then, Australia in general pursued protectionist policies.

PROTECTIONIST PARTY The Protectionist Party was one of the two main non-Labor political parties, the other being the *Free Trade Party, in the early Commonwealth Parliament. Members of the Protectionist Party held a wide range of views regarding the extent to which *protection should be practised. The Protectionist Party governed federally from 1901 to 1904, and from 1905 to 1908. Some Protectionists, led by Allan McLean, governed in coalition with the Free Traders in the *Reid-McLean Government (1904–5). Some joined with moderate Free Traders in 1906 and became known as the Tariff Reformers. The Protectionist Party was led by *Barton (1901–3) and *Deakin (1903–9). The Protectionists merged with the Free Traders and the Tariff Reformers to form the *Fusion Party in 1909. There were also some colonial (from 1901, state) protectionist parties in the late nineteenth and early twentieth centuries.

PYJAMA GIRL   The popular name given to a woman whose battered and partly-burnt body, clad in pyjamas, was found in a culvert near Albury, *New South Wales, on 1 September 1934. Her identity remained a mystery for almost ten years, during which time her body was preserved. Rewards were offered for information about her identity and that of her killer, there was widespread publicity in Australia and other countries, and many people viewed the body. In 1944 it was finally identified as being that of (Florence) Linda Agostini (née Platt). Her husband, Antonio Agostini, was charged with her murder, found not guilty of murder but guilty of manslaughter, and sentenced to six years' gaol with hard labour. He served three years and nine months in Pentridge Prison, before being deported to his native Italy in 1948. Agostini died in Sardinia in 1969.

QANTAS

*One of the first two aircraft operated by the Queensland and Northern Territory Aerial Services Limited (**Qantas**), a two-passenger Avro 504K, a war surplus bi-plane with a speed of 105 kilometres per hour. (Qantas Airways Ltd)*

QANTAS is Australia's international airline. Two Australian Flying Corps veterans of the First World War, Lieutenants P. J. McGinness and W. Hudson Fysh, grazier Fergus McMaster and others founded Queensland and Northern Territory Aerial Services Limited in 1920. Its headquarters were at Winton, Queensland, and it operated a passenger, mail, and cargo service in western Queensland and the Northern Territory. In 1934 Qantas and Imperial Airways Limited founded its successor, Qantas Empire Airways Limited, which flew the southern leg of the route to Britain. This was Australia's first international air service. The headquarters were transferred to Sydney in 1938, and Qantas expanded its international services. The Commonwealth Government nationalized Qantas in 1947. Its name was changed to Qantas Airways Limited in 1967. Qantas is said to be the second oldest international airline in the world.

QUEENSLAND, a state of the *Commonwealth of Australia, was known as the *Moreton Bay District in its early years of White settlement, when it formed part of New South Wales. It was separated from New South Wales, renamed Queensland, and granted *responsible government in 1859. It became a state of the newly created Commonwealth of Australia in 1901. Cattle production, the sugar industry (which for many years relied heavily upon the labour of Pacific Islanders brought to Australia during the *blackbirding period) and mining (especially since the 1950s and 1960s) have been particularly significant in Queensland's history.

Queensland and Northern Territory Aerial Services *Qantas

Queenstown, a town in western Tasmania, owes its existence to copper mining at nearby Mount Lyell, 6 kilometres to the north-east, where gold was discovered in the early 1880s followed by discoveries of copper and silver. Queenstown, named after Queen Victoria, was Tasmania's third largest town in 1901, with a population of over five thousand. It became a municipality in 1907. Logging, smelter fumes, and erosion largely denuded the hills around Queenstown.

R

RAAF *Royal Australian Air Force

Radio Australia *Australian Broadcasting Commission

Raffles Bay *Northern Territory; *Port Essington

Railways began to be built in Australia during the 1850s. The first public steam railways to open in each of the colonies were those between Melbourne and present-day Port Melbourne in Victoria in 1854: Redfern and Granville in New South Wales in 1855: Adelaide and Port Adelaide in South Australia in 1856; Ipswich and present-day Grandchester in Queensland in 1863; *Launceston and Deloraine in Tasmania in 1871; and Geraldton and Northampton in Western Australia in 1879. The Northern Territory's first railway, between Darwin and Pine Creek, opened in 1889. Some lines were built by private companies, but colonial governments soon took responsibility for most railways (and the Commonwealth Government also became involved from 1911). The length of railway lines increased from 243 miles (391 kilometres) in 1861 to more than 10 000 miles (16 100 kilometres) in 1890, and further lines were built during the twentieth century. (The opening in 1917 of the transcontinental railway from Port Augusta in South Australia to *Kalgoorlie in Western Australia, built as a condition of Western Australia joining the Commonwealth, was a particularly notable event.) The use of gauges of three different widths (3 feet 6 inches, 4 feet 8½ inches — standard gauge — and 5 feet 3 inches) in various parts of the country caused numerous problems. Some lines have since been converted to standard gauge, and additional standard gauge lines have been built. Diesel and electric power replaced steam during the 1950s and 1960s.

RAN *Royal Australian Navy

Rats of Tobruk Popular name given to the Allied servicemen who defended Tobruk, Libya, while it was under siege by the Axis Powers in 1941, during the Second World War. 'Lord Haw-Haw' (William Joyce) coined the term; during a German radio broadcast he described the men as 'the Rats of Tobruk, those self-supporting prisoners of war'. He meant the term to be derogatory. Instead, the men adopted it as their name. Many Australians were involved during the siege; more than 800 Australians were killed, more than 2200 were wounded, and about 700 were captured.

Redfern, William (1774?–1833), surgeon, was probably born in Canada, but grew up in England. He was commissioned surgeon's mate in HMS *Standard*, the crew of which was involved in the mutiny of the fleet at the Nore in 1797. Redfern was court-martialled and sentenced to death. He was reprieved, spent four years in jail, and was then transported to New South Wales. He was assistant surgeon at *Norfolk Island from 1802 to 1808, and *King granted him a free pardon in 1803. Redfern was assistant surgeon in New South Wales from 1808 to 1819, when he resigned because he was not appointed as principal surgeon. His report on mortality rates on three convict ships, written in 1814, led to significant reforms. As well as his hospital work, he had the colony's largest private medical practice. Redfern was also active in public affairs. *Macquarie appointed him a magistrate in 1819, but Bathurst, the British Secretary of State for the Colonies, vetoed the

*Many **railways** were built in the Australian colonies during the second half of the nineteenth century. This illustration,* First rail excursion, departing Launceston to the 6½ mile peg, *from the* Illustrated Australian News, *4 September 1869, shows the opening of the first section of the line between Launceston and Deloraine in Tasmania. (National Library of Australia)*

appointment in 1820. In 1821 Redfern sailed to England, where he and Edward Eager successfully petitioned for the lifting of certain restrictions imposed upon *emancipists. In 1824 he returned to New South Wales, where he concentrated more on farming and less on his practice, which he gave up in 1826. Redfern went to Edinburgh in 1828, and died there.

REIBEY (née Haydock), MARY (1777–1855), business woman, was born in England. She was convicted of horse-stealing and sentenced to seven years' *transportation in 1790. After her arrival in Sydney in 1792, she was assigned to *Grose, and in 1794 she married (Sydney) Thomas Reibey, an Irish free settler. (The surname has also been spelt Reiby, Raby, and Rabey.) Mary Reibey bore seven children and assisted her husband with his business interests. After his death in 1811 she maintained and expanded those interests. She and two of her daughters visited England from 1820 to 1821. Mary Reibey was also active in colonial public affairs, notably those relating to religion, education and charity. She died at Newtown, Sydney.

REID, SIR GEORGE HOUSTON (1845–1918), politician, was born in Scotland, the son of a Presbyterian minister. Reid's family migrated to Melbourne in 1852, and moved to Sydney in 1858. Reid left school at the age of thirteen and worked as a clerk. After studying law, he was admitted to the Bar in 1879. He was a Free Trade Member of the Legislative Assembly in New South Wales from 1880 to 1884, and again from 1885 to 1901, and was Premier and Treasurer from 1894 to 1899. He implemented economies in government spending, introduced land tax, and established a 'non-political' public service board. Reid, who had earlier been dubbed 'Yes-No' Reid because of his attitude towards federation, was a Member of the House of Representatives from 1901 to 1909. He led the Free Traders in the first Commonwealth Parliament, then became Prime Minister and Minister for External Affairs, leading a coalition of Free Traders and Protectionists in 1904–5. After the defeat of his Government (sometimes known as the Reid–McLean Government) he continued to lead the Opposition until his retirement as Free Trade Leader in 1908. Reid became the first Australian High Commissioner in London in 1910, holding this position until 1916, in which year he was elected to the British House of Commons. He died in London.

REMITTANCE MEN were immigrants in Australian and other colonies during the nineteenth century who survived on remittances from home (usually Britain and Europe), rather than wages. In some cases, their families paid them to stay abroad. They were often looked upon with scorn.

RESERVE BANK OF AUSTRALIA Australia's central bank. In 1959 the Commonwealth Bank of Australia was divided into the *Commonwealth Banking Corporation and the Reserve Bank of Australia. The Reserve Bank's main functions, according to the *Reserve Bank Act* 1959, are to 'contribute to the stability of the currency of Australia; the maintenance of full employment in Australia; and the economic prosperity and welfare of the people of Australia'.

RESPONSIBLE GOVERNMENT A form of self-government. Responsible government was proclaimed in New South Wales, Victoria and Tasmania in 1855, in South Australia in 1856, in Queensland in 1859, and in Western Australia in 1890. The colonial constitutions established bicameral *parliaments, and these continue in the states, except Queensland, which abolished its Legislative Council in 1922.

RETURNED SERVICES LEAGUE OF AUSTRALIA (RSL) A pressure group representing the interests of ex-servicemen and ex-servicewomen. The largest of a number of similar Australian organizations, it was formed in 1916, during the First World War, as the Returned Sailors' and Soldiers' Imperial League of Australia. Its name was changed in 1940, during the Second World War, to the Returned Sailors', Soldiers' and Airmen's Imperial League of Australia. It was changed again, to the above name, in 1965. The organization has worked mainly in areas such as repatriation and welfare. Although it claims to be non-political, it has spoken out on topics such as defence and foreign policy. Before 1982 members needed to have served in a theatre of war; after then, anyone who served in the regular defence forces for six months or more became eligible. There were about 250 000 members Australia-wide in 1992.

RFDS *Royal Flying Doctor Service of Australia

RICHARDSON, ETHEL FLORENCE LINDESAY *Robertson, Ethel Florence Lindesay

'RICHARDSON, HENRY HANDEL' *Robertson, Ethel Florence Lindesay

Unsigned and undated portrait of Mary **Reibey**. (Mitchell Library, State Library of New South Wales)

George **Reid**, seen here in a photograph taken in the 1890s, was Australia's Prime Minister from 1904 to 1905. (Australian Natives' Association)

'RIDDLE OF THE RIVERS' *Sturt, Charles

RIDLEY'S STRIPPER, a mechanical wheat harvester, is considered to have been one of the most significant inventions in the development of farming in Australia. The first such machine was apparently made by John Ridley (hence its popular name), and was first used in South Australia in 1843. Another South Australian, J. W. Bull, had earlier suggested a similar idea.

RIPON REGULATIONS Land regulations for the Australian colonies, introduced in 1831 by the British Secretary of State for the Colonies, Viscount Goderich (created Earl of Ripon in 1833). They replaced the existing system of disposal of Crown land, a combination of sales and grants, with a system of sales (except in cases where grants had already been promised). Henceforth, Crown land was to be sold at auction at a minimum price of five shillings an acre. The proceeds of the land sales were to be used to assist the *immigration of free settlers.

ROBERTS, THOMAS WILLIAM (1856–1931) painter, was born in England, and came to Melbourne with his widowed mother in 1869. He worked as a photographer's assistant, and later drew for newspapers and journals. He studied at night, first at the Collingwood Artisans School of Design, and then at the National Gallery School. In 1881 he went to England, where he continued to draw for newspapers, and studied at the Royal Academy School. He and several friends went on a walking tour of Spain in 1883. He returned to Australia in 1885. Tom Roberts was largely responsible for founding the *Heidelberg School, and established, with the group's other members, artists' camps in Box Hill and Heidelberg in the 1880s. Roberts and *Streeton then established a camp at Mosman, Sydney, in 1891. Roberts also taught and painted portraits in Melbourne and later in Sydney. He became the foundation president of the New South Wales Society of Artists in 1895. Roberts lived abroad, mainly in England, from 1903, and continued to paint. During the First World War he worked as a hospital orderly in England. He returned permanently to Australia in 1923. His paintings include *Bourke Street* (1885–6), *Coming South* (1886), *Shearing the Rams* (1889–90), *The Breakaway* (1891), and *Opening of the First Commonwealth Parliament* (1901–3).

ROBERTSON (née Richardson), ETHEL FLORENCE LINDESAY (Henry Handel Richardson) (1870–1946), novelist, more commonly known by her pseudonym, was born in East Melbourne. Her Irish-born father was a medical doctor; her mother later became a postmistress. She was educated at the Presbyterian Ladies College, Melbourne, and the Leipzig Conservatorium. After leaving Australia for Leipzig in 1887, she spent most of her life in England, returning only once to Australia, in 1912, while researching *The Fortunes of Richard Mahony*. She is best known for this trilogy, which was first published separately as *The Fortunes of Richard Mahony* (usually known as *Australia Felix*, part of its half-title) (1917), *The Way Home* (1925) and *Ultima Thule* (1930). The trilogy, set largely in colonial Australia, traces the decline of Richard Mahony, a character modelled upon the author's own father. Her other novels and stories include

ROBERTSON, SIR JOHN

*Tom **Roberts**'s works include numerous portraits, such as this one entitled* Mrs L. A. Abrahams *(1888). (National Gallery of Victoria)*

Maurice Guest (1908), *The Getting of Wisdom* (1910) and *The End of a Childhood and Other Stories* (1934). Her *Myself When Young* (1948) is an unfinished autobiography.

ROBERTSON, SIR JOHN (1816–91), politician, was born in England, and came to New South Wales with his family in 1822. He became a *squatter in the 1830s. Robertson entered New South Wales politics, and was a Member of the Legislative Assembly for most of the period from 1856 to 1886 (and a Member of the Legislative Council twice during that time). He was the Premier of New South Wales five times (1860–1, 1868–70, 1875–7, 1877, 1885–6) and held various other portfolios. Robertson was largely responsible for the introduction of land reforms in New South Wales in 1861.

ROBINSON, GEORGE AUGUSTUS (1788–1866), protector of *Aborigines, was born in England. His father was a builder. Robinson received little formal education, and worked for the Engineers Department at Chatham before migrating to Australia in 1824. He worked as a builder in Hobart. In 1829 he was appointed to work with Aborigines. For the next few years, Robinson travelled around *Van Diemen's Land, making contact with Aborigines and persuading them to live in captivity. Those persuaded were taken to a series of settlements on Bass Strait islands. From 1835 to 1839 Robinson himself took charge of the final such settlement, on Flinders Island. In 1838 he became Chief Protector of the newly formed *Port Phillip Protectorate, and held that position until the abolition of the Protectorate in 1849. In 1852 he returned to England, living there and in Europe for the rest of his life. He died at Bath, England. Some of his writings were later published in *Friendly Mission: The Tasmanian Journals and Papers of George Augustus Robinson 1829–1834* (1966), edited by N. J. B. Plomley.

ROCKS, THE An area on the western side of Sydney Cove, which takes its name from the large sandstone rocks found there. Makeshift buildings were erected there as early as 1788, and most of the early buildings were for maritime, defence, and commercial purposes. During the nineteenth century the Rocks gained a notorious reputation for harbouring groups of criminals, prostitutes, and *larrikins. Many of the nineteenth-century buildings at the Rocks have been preserved and restored, and the area has become a tourist attraction.

ROSE HILL *Parramatta

ROWAN (née Ryan), **MARIAN ELLIS** (1848–1922), artist, naturalist, and explorer, was born in Melbourne, and was educated at Miss Murphy's High School for Girls in Brighton, but received no formal art training. She travelled extensively throughout the Australian colonies, New Zealand, the USA, New Guinea, and elsewhere, painting birds, wildflowers, insects, and butterflies. Baron Ferdinand von

John Eyre's view of Sydney from the west side of the cove, showing part of the Rocks area.

*Mueller used some of her work to classify previously unidentified Australian flowers. Ellis Rowan successfully exhibited her work in various countries and gained an international reputation. She died at Macedon, Victoria. Her publications include *A Flower Hunter in Queensland and New Zealand* (1898). The National Library of Australia, Canberra, holds some 950 of Rowan's studies of Australian and New Guinea flora and fauna, now generally considered of greater scientific than artistic value.

ROYAL AUSTRALIAN AIR FORCE (RAAF) The most recently established of the Australian armed services, the others being the *Royal Australian Navy and the *army. The Australian Flying Corps (AFC) was formed in 1915, during the First World War. Its members served in the Middle East and on the Western Front. Some formed part of the AIF, others part of the British Royal Flying Corps. The AFC was disbanded after the war. Its successor, the RAAF, was established in 1921. Early development was slow, but it expanded greatly during the Second World War. Australia participated in the Empire Air Training Scheme, the purpose of which was to share the dominions' resources to train airmen, which began in 1939. RAAF members served in Europe, the Middle East, Asia, and the Pacific during the war (some of the them as part of the British Royal Air Force). Women were first accepted in the air force as nurses in 1940, and in some other positions in 1941. The women's auxiliary, but not the nursing service, was disbanded in 1947. Women were again accepted in the air force from 1950. Since the Second World War, RAAF members have served overseas in the *Korean War, the *Malayan Emergency, the *Vietnam War, and several peacekeeping operations.

ROYAL AUSTRALIAN NAVY (RAN) One of Australia's three armed services, the others being the *army and the *Royal Australian Air Force. The Royal Australian Navy was formally established in 1911. Previously, the Australian

Ellis Rowan's Unidentified Australian Flowers, one of her many works held by the National Library of Australia. (National Library of Australia)

ROYAL AUSTRALIAN NAVY

*Members of the **Royal Australian Navy** participated in the Gulf War. Ron Tandberg's view of their role,* I don't want to go . . . but someone's got to protect little countries if they've got lots of oil, *was published in the* Age *on 31 December 1990. (La Trobe Collection, State Library of Victoria)*

colonies had depended largely upon Britain for naval defence, paying subsidies for this purpose from 1887. Small colonial naval forces had also been raised during the second half of the nineteenth century. A naval contingent took part in the *Boxer Rising. The Commonwealth Government took responsibility for the various local forces following federation in 1901. Australian ships, placed under British Admiralty command, were largely engaged in convoy work during the First World War. Naval forces were involved in the capture of German New Guinea, and the *Sydney* destroyed the German raider *Emden* at the Cocos Islands in November 1914.

Two Australian submarines were destroyed during the war; the *AE1* was lost near New Guinea in 1914, and the *AE2* was sunk in the Sea of Marmara in 1915. During the Second World War, Australian ships served in many parts of the world, and were engaged in much action. The sinking of the Italian cruiser *Bartolomeo Colleoni* by the *Sydney* off Crete in July 1940 was particularly notable. More than thirty Australian vessels were lost during the war, including the *Sydney*, which was sunk off Western Australia in November 1941. The Royal Australian Navy was also involved in the *Korean War, the *Vietnam War, and the Gulf War, and has taken part in a number of peace-keeping operations. Women were first accepted in the navy in 1941.

ROYAL FLYING DOCTOR SERVICE OF AUSTRALIA (RFDS) An organization that provides a medical service, using radio and aeroplanes, for outback areas. The Reverend John Flynn was largely responsible for the foundation of the Aerial Medical Service in 1928. Based at Cloncurry in Queensland, it was founded under the auspices of the Presbyterian Church's Australian Inland Mission, of which Flynn was founder and superintendent. Responsibility for the organization was transferred from the Australian Inland Mission to the states in 1933, and it became known as the National Aerial Medical Service. In 1942 its name was changed to the Flying Doctor Service of Australia, the word 'Royal' being added to its title in 1954. The organization receives public and private funding, and now has bases in many parts of Australia. Its radio network has also been used for the Schools of the Air (an educational service), and for sending telegrams.

RSL *Returned Services League of Australia

'RUDD, STEELE' *Davis, Arthur Hoey

RUM CORPS *New South Wales Corps

RUM JUNGLE, in the Northern Territory, was the site of Australia's first large uranium project. Uranium was found there in 1949, and production began in the early 1950s. It had ceased by 1971.

RUM REBELLION Popular name given to the overthrow in 1808 of the Governor of New South Wales, *Bligh, by leading colonists and officers of the *New South Wales Corps. The Corps has become known as the Rum Corps because of its members' involvement in the rum trade. Background causes of the Rum Rebellion included the personalities of some of those involved, especially Bligh and *Macarthur, and Bligh's successful attempts to limit the powers of the rum traders. Immediate causes included the last of a series of legal cases involving Macarthur. George Johnston led the Corps to arrest Bligh on 26 January. He dubbed himself 'Lieutenant-Governor' and controlled the colony until July, when *Foveaux took control. William *Paterson later took over from Foveaux and remained in control until the next governor, *Macquarie, took office in 1810. Macarthur was Colonial Secretary during the period when Johnston controlled the colony. Bligh was confined in Sydney for some time, then sailed to Van Diemen's Land in 1809, returned to Sydney in 1810, and finally sailed for England that year. Johnston returned to England, where he faced a court martial; he was found guilty and cashiered in 1811. Bligh was virtually exonerated at Johnston's court martial. Macarthur returned to England, where he remained until 1817, aware that he risked arrest if he returned to the colony. The New South Wales Corps was recalled.

RYAN, THOMAS JOSEPH (1876–1921), politician, was born at Port Fairy, Victoria, and was educated at Port Fairy State School, Xavier College, Melbourne, and the University of Melbourne. Ryan, whose father was a

*Sketch (c. 1808) by an unknown artist of Governor Bligh's arrest during the **Rum Rebellion**. Whether Bligh was really under the bed at the time has since been the subject of debate. (Mitchell Library, State Library of New South Wales)*

grazier, taught at schools in several states, and practised as a lawyer in Queensland, before serving as a Labor Member of the Queensland Legislative Assembly from 1909 to 1919 (becoming parliamentary leader of his party in 1912). He was Premier, Chief Secretary and Attorney-General from 1915 to 1919. His Government was responsible for much industrial legislation, including Acts dealing with arbitration and workers' compensation; established a number of state enterprises, among which were butchers' shops, station properties, an insurance office, and a hotel; and unsuccessfully attempted to abolish the Legislative Council. Ryan's strong stand against *conscription during the First World War brought him into conflict with the Prime Minister, *Hughes. Ryan was a Labor Member of the House of Representatives for West Sydney from 1919 until he died. He was Deputy Leader of the federal parliamentary Labor Party from 1920.

S

SAVERY, HENRY (1791–1842), writer, was born in England, and became a sugar broker in Bristol. He went bankrupt, was convicted of forgery, and was transported for life, arriving in 1825 in Hobart, where he worked in government offices. While imprisoned for debt, Savery wrote a series of sketches about life in Hobart, which were published in the *Colonial Times* during 1829 under the pseudonym of 'Simon Stukeley'. The sketches were published anonymously in book form as *The Hermit in Van Diemen's Land* (1829), said to be the first volume of Australian essays, and became the subject of a libel action in 1830. Savery also wrote the first Australian novel, the partly autobiographical *Quintus Servinton* (1830–1), which was published anonymously in three volumes. He became a farmer, was granted a *ticket of leave in 1832 (withdrawn for some time in 1833), and received a conditional pardon in 1838. Savery became insolvent again in 1839, was convicted of forgery a second time in 1840, and was sent to *Port Arthur, where he died.

SCHOOLS OF ARTS *Mechanics Institutes

SCOTT, ROSE (1847–1925), feminist and social reformer, was born in New South Wales. She is perhaps best known for her work in the movement for women's *suffrage. She and others founded the Womanhood Suffrage League in Sydney in 1891, and she remained secretary of the League until it was dissolved in 1902, after women in New South Wales had gained the vote. Rose Scott then founded, and was president of, the Women's Political Educational League, which existed until 1910. She was also involved in other organizations, advocating peace and women's rights, and she campaigned for legislative reforms in areas relating to women and children. She died in Sydney.

SCOTTISH MARTYRS Political prisoners transported to New South Wales after celebrated Scottish sedition trials in 1793 and 1794. They were Thomas Muir, Thomas Fyshe Palmer, William Skirving, Maurice Margarot, and Joseph Gerrald. All had been active in organizations, influenced by the French Revolution, advocating political reform. Palmer was sentenced to seven years' *transportation, the others to fourteen years'. All but Gerrald arrived in New South Wales in 1794; he arrived the following year, and died of tuberculosis in 1796. As political prisoners, they were excused from compulsory labour. Muir farmed until his escape on an American ship, the *Otter*, in 1796. After overcoming extreme difficulties, including injury and imprisonment, he reached Paris in 1797. He died at Chantilly, near Paris, in 1799. Palmer farmed and traded until he sailed for England in 1801. He reached Guam, where he died in 1802. Skirving farmed until his death in 1796. Margarot continued his political activities; he is said to have helped to plan the *Castle Hill Uprising of 1804. He was sent to *Norfolk Island, *Van Diemen's Land, and *Newcastle. In 1810 he returned to England, where he died in 1815.

SCULLIN, JAMES HENRY (1876–1953), politician, was born in Victoria, was educated at Mt Rowan State School (later attending night classes), and worked variously as a gold-miner, shopkeeper, and organizer for the *Australian Workers' Union. He was a Labor Member of the House of Representatives from 1910 to 1913, and again

J. H. Scullin was Australia's Prime Minister from 1929 to 1932. (National Library of Australia)

from 1922 to 1949. After losing his seat in 1913, Scullin worked as a journalist and edited the *Ballarat Evening Echo*. He led the Opposition in 1928–9, and then was Prime Minister and Minister for External Affairs from 1929 to January 1932. As Prime Minister, he was confronted with deteriorating economic conditions and deepening divisions within his own party. His Government greatly increased tariffs in an unsuccessful attempt to reduce unemployment. In 1931 he appointed the first Australian Governor-General, *Isaacs, despite opposition from King George V. After the defeat of the Labor Government, Scullin led the Opposition from 1932 to 1935. In 1935 he resigned as federal parliamentary leader of the Labor Party, but did not retire from the Commonwealth Parliament until 1949.

SEALERS from Australia, Britain, America and elsewhere operated around Australia from the 1790s until about the 1830s. Seals were hunted for their skins and oil. The Bass Strait islands and *Kangaroo Island were important sealing grounds during the early years. Sealers then moved to places such as New Zealand and Macquarie Island. Whale and seal products were Australia's main exports until the 1830s. Fur seals were almost extinct by the late 1850s.

SEATO *South-East Asia Treaty Organization

SECESSION MOVEMENT A significant secession attempt by *Western Australia during the *Great Depression. Western Australia wished to secede from the *Commonwealth of Australia, becoming instead a self-governing member of the *Commonwealth of Nations. A Dominion League of Western Australia,

*The main body behind the Western Australian **secession movement** of the 1930s was the Dominion League of Western Australia. This photograph was taken at their Grand Victory Festival in 1933. (National Library of Australia)*

241

SECOND FLEET

*Enlistment poster from the **Second World War**,* What's your excuse for not joining your comrades in the AIF? *(Tasmaniana Library, State Library of Tasmania)*

advocating secession, was formed in 1930. In 1933 Western Australians voted two-to-one in favour of secession at a referendum, at which voting was compulsory, and a deputation petitioned the British Parliament in 1934. The petition was rejected the following year.

The main reasons for the movement were emotional and financial (especially increasing dissatisfaction with the Commonwealth Government's *protection policy).

SECOND FLEET Popular name for the

second group of ships bringing convicts and supplies from England to New South Wales, in 1790. (The *Lady Juliana*, carrying about 220 female convicts, sailed from England in July 1789, between the *First Fleet and the Second Fleet, and arrived in early June 1790.) The *Surprize*, the *Neptune*, and the *Scarborough*, carrying about 1000 convicts, left England together on 19 January 1790, and arrived in late June of that year. The *Justinian*, a store ship, went with them. Conditions on board were grim, and about a quarter of the convicts died during the voyage. Officers and men of the *New South Wales Corps sailed with the Second Fleet.

SECOND WORLD WAR (1939–45) A major war, in which many countries were involved, fought between the Allied Powers and the Axis Powers. The Allied Powers included Britain (1939–45), the USA (1941–5), France (1939–40), China (1941–5), and Australia (1939–45). The Axis Powers included Germany (1939–45), Italy (1940–3, but a 'co-belligerent' with the Allies from 1943 onwards), and Japan (1941–5). Dozens of other countries were also involved. The long-term causes, which were complex and numerous, included the inadequacy of the peace settlement reached after the *First World War, the rise of Nazism in Germany, and the rise of Fascism in Italy and elsewhere. The short-term cause was the German invasion of Poland on 1 September 1939. Britain declared war on Germany two days later, and other countries soon became involved. The main theatres of war were in Europe, North Africa, and the Pacific. Germany surrendered on 7 May 1945. Japan, following the dropping of atomic bombs on Hiroshima on 6 August 1945, and on Nagasaki on 9 August 1945, agreed to accept Allied terms on 14 August, and formally surrendered on 2 September. Australia, as a member of the British *Commonwealth of Nations, automatically followed Britain to war on 3 September 1939, as she had at the beginning of the First World War. Australia declared war on Italy in 1940, Japan, Finland, Hungary, Rumania, and Bulgaria in 1941, and Thailand in 1942. The *Royal Australian Navy, the Australian *army, and the *Royal Australian Air Force all participated in the war. The gross enlistments during the war were 45 800 men and 3100 women in the Royal Australian Navy; 691 400 men and 35 800 women in the army; and 189 700 men and 27 200 women in the Royal Australian Air Force. Australia's main expeditionary force was the second *Australian Imperial Force, whose members were all volunteers. From 1940 onwards, members of the second AIF served in the Middle East, where some became the now-famous *Rats of Tobruk. Some also fought in Greece and Crete. Other members of the second AIF went to Malaya in 1941; many of those were taken prisoner by the Japanese following the fall of Singapore on 15 February 1942. Some others remained in Australia. The Australian troops in the Middle East returned to the Pacific in 1942 and 1943, following Japan's entry into the war on 7 December 1941. Second AIF troops served in the Pacific region (including Australia) for the remainder of the war. Militia forces, whose members were conscripts, served in Australia, New Guinea, and some other islands. The RAAF and RAN served in many parts of the world. Australian battle casualties (including those of prisoners-of-war) included 1900 deaths in the RAN, 6460 in the RAAF, and 18 713 in the army. More than 30 000 Australians were taken prisoners-of-war, mainly by the Japanese. The *Menzies United Australia Party Government reintroduced military *conscription for home service in 1939. The *Curtin Labor Government introduced a limited form of military conscription for overseas service in 1943. A form of industrial conscription, for men and women, was also introduced. Women were eventually accepted in the armed services, albeit in restricted roles, during the war, and an Australian Women's Land Army was formed. Various volunteer home-defence forces were raised. The Commonwealth Government, whose powers were increased during the war, notably through the introduction of *uniform taxation, had introduced wartime legislation such as the *National Security Act* 1939. However, the later *'Fourteen Powers' referendum failed. The *Communist Party of Australia was banned for some time. The *Australia First Movement was suppressed. More than 25 000 German, Italian, and Japanese prisoners-of-war were held in Australia during the war. A large number of Japanese prisoners attempted to escape in the *Cowra outbreak. Rationing of petrol, clothing, food, and other goods and services was introduced. The Japanese launched air raids on *Darwin on 19 February 1942. In the following months there were further air raids on Darwin, Wyndham, Broome, Derby, and elsewhere. There were hundreds of casualties and some property damage. Three Japanese midget submarines attacked Sydney Harbour on 31 May 1942, also causing some casualties and property damage. Submarines shelled Sydney and Newcastle on 7 June 1942, but with little effect. The *Kokoda Trail was the closest the Japanese forces came to Australia by land. There was controversy over the alleged existence of the *Brisbane Line, beyond which Australia would not be defended. Following Japan's entry into the war, Australia turned increasingly towards the USA for support. The latter had already been providing *Lend-Lease aid to Australia. Americans began to arrive in Australia from late 1941. General Douglas MacArthur, who went on to command the Allied operations in the south-west Pacific, arrived in Australia in March 1942. The Battle of *Brisbane has become part of Australia's folklore. Australia's increased reliance upon the USA was later reflected in the signing of the *ANZUS Pact after the war.

SECRET BALLOT A method whereby electors cast their votes in secret. It is known also as the Australian ballot. Australia led the world in introducing the secret ballot for political elections. Victoria and South Australia first introduced provision for it (for both houses) in 1856, Tasmania (for both houses) in 1858, New South Wales (for the lower house) in 1858, Queensland (for the lower house) in 1859, and Western Australia (for the upper house) in 1877. The secret ballot has always been used for elections for the Commonwealth House of Representatives and Senate.

SELECTORS Popular name given to a large number of small farmers in Australia during the second half of the nineteenth century. By the 1850s *squatters had acquired most of the land suitable for agricultural and pastoral purposes. Increasing demands were made, especially by those who had come to the colonies during the gold-rushes, for land to be made available at low cost for small farms. All of the colonies introduced systems providing for farms of limited size and the sale of land on

*Life was much harsher for many **selectors** than Julian Ashton's* The Selector's Daughter *(1907) might suggest. (National Gallery of Victoria)*

credit. Selectors generally were not very successful in New South Wales, but fared better in Victoria and South Australia. There were few selectors in Tasmania, and Queensland and Western Australia remained largely pastoral colonies. Factors leading to the failure of many selectors included the unsuitability of much Australian land for agriculture, the use of devious and obstructive tactics by some squatters, and many selectors' ignorance of farming and lack of capital.

SHEARERS' STRIKES Major strikes in Queensland and New South Wales in 1891 and 1894. Shearers were fighting for the principles of unionism and the 'closed shop' ('an establishment in which only trade union members are employed', or 'a trade restricted to members of a [particular] trade union'). Pastoralists were fighting for 'freedom of contract' (the right to employ anyone). The strikes were marked by violence and bitterness on both sides. Non-union labour was used. Union leaders, including some from the now-famous *Barcaldine shearers' camp of 1891, were arrested on charges such as conspiracy, seditious language, and riot. Some were jailed. The unions were defeated.

SHEFFIELD SHIELD *Cricket

SHENANDOAH A ship whose visit to Melbourne during the American Civil War led to a diplomatic incident. The *Shenandoah*, an American Confederate ship, visited Melbourne from 25 January to 18 February 1865, against the wishes of the United States Consul. Her captain, James Waddell, with permission from the Victoria Governor, Sir Charles Darling, had repairs made and bought supplies; without permission, he also recruited sailors. After the American Civil War ended, the American Government made the British Government pay compensation for damage done to Union ships by the *Shenandoah* after she left Melbourne.

'**SILVER CITY**' *Broken Hill

SIMPSON, JOHN *Kirkpatrick, John Simpson

SMALLPOX A highly contagious and usually fatal disease. There appear to

*The fêting of the officers of the Confederate ship **Shenandoah** when they visited Victoria in 1865 is reflected in this contemporary illustration of a ball held in their honour at Ballarat. (La Trobe Collection, State Library of Victoria)*

have been smallpox epidemics among *Aborigines, in at least the eastern part of Australia, in 1789 and between about 1829 and 1831. There is debate about the way in which the disease spread. There was also an outbreak of smallpox among White settlers in Australia in the 1880s.

SMITH, SIR KEITH MACPHERSON (1890–1955) and SIR ROSS MACPHERSON (1892–1922), aviators, were brothers, born in Adelaide and educated there and in Scotland. During the First World War Ross served with the AIF and the Australian Flying Corps, and Keith with the Royal Flying Corps. After the war they made the first flight from England to Australia. In 1919 the Commonwealth Government offered a prize of £10 000 for the first Australians to make such a flight in thirty days or less. Ross (pilot), Keith (co-pilot and navigator) and two mechanics, J. M. Bennett and W. H. Shiers, left Hounslow, on 12 November 1919. They arrived in Darwin twenty-seven days and twenty hours later. Frank *Hurley flew on the last leg of the flight. The Smith brothers won the prize and were knighted. Sir Ross Smith and Bennett were killed while testing an amphibian aircraft in preparation for another long-distance flight. Sir Keith Smith, who later chaired the Australian branch of Vickers Limited, died in Sydney.

SNOWY MOUNTAINS HYDRO-ELECTRIC SCHEME A major engineering feat. The Scheme generates electricity for the Australian Capital Territory, New South Wales, and Victoria, and also provides water for the Murray and Murrumbidgee irrigation areas. It involves the diversion of water from the Snowy River through tunnels in the Snowy Mountains, in south-eastern New South Wales. Its construction, almost entirely funded by the Commonwealth Government, began in 1949 and ended in 1972.

SOCIAL CREDIT A theory of economic reform propounded by a Scottish engineer named Clifford H. Douglas, and so known also as Douglas Credit. Although Douglas's ideas circulated in Australia during the 1920s, they were not influential until the *Great Depression. Social Credit organizations were formed from 1930 onwards; Social Credit newspapers, such as the *New Era*, *Freedom* (later *New Economics*), *New Times*, and *New World* were published; Douglas himself visited Australia in 1934; Social Credit political parties, such as the Douglas Credit Party, were formed. No official Social Credit candidates were ever elected to parliament. The Social Credit movement had virtually dissolved in Australia by 1940, except in Queensland, where it continued until the 1960s. Its supporters have mainly been absorbed into other organizations.

'SOCIALIST TIGER' During the federal election campaign of 1906, *Reid, leader of the Free Traders, referred to the *Australian Labor Party as a 'socialist tiger'. The slogan formed part of a broader 'anti-socialist' campaign that Reid had begun in 1904.

SOLDIER SETTLEMENT A land policy implemented after each of the two world wars. It was a form of *closer settlement designed to establish returned soldiers on small farms, often in previously unsettled regions. Governments gave soldier settlers credit and other assistance. There were about 37 000 soldier settlers after the First World War, mainly along the Murray and Murrumbidgee Rivers, in the wheat-growing regions of South Australia and Western Australia, and in the Mallee in Victoria. They frequently faced problems such as lack of experience and capital, farms that were too small, and unproductive land. Many left their farms during the 1920s and 1930s, and the Commonwealth Government had to cover many of their debts. Soldier settlement was implemented on a smaller scale after the Second World War.

SORELL, WILLIAM (1775–1848), soldier and colonial administrator, was probably born in the West Indies, where he later served during his military career. He served in other places, including the Cape of Good Hope, where he demonstrated his administrative ability, before resigning from the army in 1813. He replaced Thomas Davey as Lieutenant-Governor of *Van Diemen's Land in 1817. Sorell reduced bushranging in the colony; introduced numerous administrative reforms; established a *penal settlement at *Macquarie Harbour and built convict barracks in Hobart; encouraged the development of the wool and wheat industries; and helped to found the Bank of Van Diemen's Land. After *Arthur succeeded him in 1824, Sorell returned to England. He did not hold administrative office again.

SORRENTO is a small town on the eastern shore of Port Phillip Bay. The British, motivated partly by fears that the French might settle in the area, established a convict settlement, the first White settlement in present-day Victoria, near the site of Sorrento in 1803. About five hundred settlers (convicts, troops, and free settlers), under the command of *Collins, arrived from England in October of that year. The settlement soon faced problems, including reluctance on the part of settlers, lack of fresh water, and heat. The leaders were also worried about the difficulties of access by sea. Collins withdrew the party and established a new settlement in *Van Diemen's Land early in 1804. The last of the party left the area in May 1804, except *Buckley, an escaped convict.

SOUTH AFRICAN WAR *Boer War

SOUTH AUSTRALIA   A state of the *Commonwealth of Australia since the latter's creation in 1901. Dutch sailors are the first Europeans known to have sighted its coast, which they did in 1627. Other Europeans, including *Flinders and *Baudin, later explored the coast. *Sealers began to use *Kangaroo Island in the early 1800s. *Sturt, and later Collet Barker, explored some areas inland. Settlement of the British province of South Australia, based partly upon *Wakefield's theory of *systematic colonization, began in 1836. The British Government and a Colonization Commission initially shared responsibility for the province. The first immigrants settled on Kangaroo Island, and on the mainland near present-day Glenelg and Adelaide. The first two governors of South Australia were *Hindmarsh and *Gawler. The province went bankrupt, but after Britain paid its debt, it became a conventional colony in 1842, and was granted *responsible government in 1856. Alone among the Australian colonies, South Australia never received convicts. It was Australia's most important wheat-growing state for much of the second half of the nineteenth century. South Australia annexed the *Northern Territory in 1863, and administered it from then until 1911.

SOUTH AUSTRALIAN COMPANY   A company founded by George Fife Angas and others in England (1835–6) with the purpose of encouraging investment and development in the new British province of *South Australia. The Company bought 13 770 acres (at twelve shillings per acre) and the pastoral rights to a further 220 000 acres (at ten shillings per square mile). It ran a whaling base at Kingscote on *Kangaroo Island from 1836 to 1841. As it grew, its activities came to include land speculation, trading, banking, and shipping. In 1949 its activities in South Australia ended.

SOUTH-EAST ASIA TREATY ORGANIZATION (SEATO) An organization established under the South-East Asia Collective Defence Treaty (the Treaty also becoming known as SEATO), which was signed at Manila in 1954. The signatories were Australia, Britain, France, New Zealand, Pakistan, the Philippines, Thailand and the USA. The Treaty, which covered South-East Asia and part of the South-West Pacific, included the following statements:

> *Each Party recognizes that aggression by means of armed attack in the Treaty Area or against any State or territory which the Parties by unanimous agreement may hereafter designate would endanger its own peace and safety, and agrees that it will in that event act to meet the common danger in accordance with its constitutional processes.*
>
> *If in the opinion of any of the Parties, the inviolability or the integrity of the territory or the sovereignty or political independence of any Party in the Treaty Area or of any State . . . is threatened in any way other than by armed attack . . . the Parties shall consult immediately in order to agree on the measures which should be taken for the common defence.*

Pakistan withdrew from the Organization in 1973, and France in 1974. The Organization itself was dissolved in 1977, although the Treaty remained in existence.

SOUTHERN CROSS FLAG *Eureka flag

'SPANISH' INFLUENZA PANDEMIC (1918–19) A pandemic (a disease 'prevalent over the whole of a country or of the world') that swept the world towards the end of the *First World War and in the immediate post-war period, causing an estimated twenty to thirty million deaths. Returning Australian soldiers, who were among the first victims, brought the disease into Australia. Protective measures were taken, such as quarantine between some of the states, closure of public places, the wearing of face masks, and inoculation. However, about twelve thousand Australians had died from the disease by the end of 1919.

SPENCE, CATHERINE HELEN (1825–1910), writer, feminist, and social reformer, was born in Scotland and migrated to South Australia with her family in 1839. She was a governess for some time. Catherine Spence wrote novels, works of non-fiction, and articles for newspapers and journals; she campaigned for various causes, including children's welfare, electoral reform, girls' education, female *suffrage, and political education of women; she became Australia's first female political candidate when she ran, unsuccessfully, for the *Federal Convention in 1897; and she preached in churches in Australia and the USA (having joined the Unitarian Church about 1856). Her publications, some of which were published anonymously, include *Clara Morison: A Tale of South Australia During the Gold Fever* (1854), which is said to be the first novel about Australia written by a woman, *Tender

Margaret Preston's portrait, Catherine Helen Spence *(1911), was painted after* Spence*'s death, but is considered to be a good likeness. (Art Gallery of South Australia)*

and True (1856), *A Plea for Democracy* (1861), calling for the introduction of Thomas Hare's system of proportional representation in South Australia, *Mr. Hogarth's Will* (1865), a textbook entitled *The Laws We Live Under* (1880), and *An Autobiography* (1910), which was completed by Jeanne Young.

SPENCE, WILLIAM GUTHRIE (1846–1926), trade unionist and politician, was born in the Orkney Islands, Scotland. His father was a stonemason. In 1852 the family migrated to Australia. Spence was not formally educated. He held various jobs, including those of shepherd, butcher boy and miner, in Victoria, and then became involved in trade union organization. He was secretary of the Creswick Miners' Union, which he helped to found in 1878; he was secretary of the Amalgamated Miners' Association from 1882 to 1891; he helped to found, and from 1886 to 1893 was the first president of, the Amalgamated Shearers' Union of Australasia; he was actively involved in the *Maritime Strike (1890) and the *Shearers' Strike (1891); in 1894 he helped to found the *Australian Workers Union, of which he was secretary from 1894 to 1898, and president from 1898 to 1917. He was a Member of the Legislative Assembly of New South Wales from 1898 to 1901. He then entered federal politics, and was a New South Wales Member of the House of Representatives from 1901 to 1917 (representing Labor, then National Labor), and then a Tasmanian Member from 1917 to 1919 (representing the Nationalists). Spence was Postmaster-General (1914–15) and Vice-President of the Executive Council (1916–17). He died at Terang, Victoria. His publications include *Australia's Awakening: Thirty Years in the Life of an Australian Agitator* (1909), *History of the A.W.U.* (1911), and various pamphlets and articles.

SQUATTERS A term with several Australian meanings. In the early nineteenth

George Lambert's The Squatter's Daughter *(1923–4).* **Squatters** *have played a significant part in Australia's history. (Australian National Gallery)*

century it was a derogatory term in New South Wales; the first squatters, many of whom were ex-convicts, occupied land without authority and often stole stock. By the 1830s the meaning had begun to change, and the many pastoralists who settled beyond the *Nineteen Counties were called squatters. Most were involved in the wool industry. In 1836 they were granted grazing rights for an annual licence fee. They demanded security of tenure and pre-emptive rights, which they gained in 1847. There were also some squatters in South Australia and Western Australia, although in much smaller numbers than in New South Wales and Victoria. Squatters gained a monopoly over much of the land in Australia that was suitable for agricultural and pastoral purposes. They became a very powerful group, socially, economically, and politically. Squatters and *selectors struggled bitterly over land during the second half of the nineteenth century. Squatters continued to be known as such even after they acquired their land freehold, and the term was eventually applied to all large pastoralists in Australia.

STATE SOCIALISM A term open to various interpretations. It can involve simply the public ownership of essential services such as gas, water, and electricity; more strongly, it can involve the communal ownership of the means of production, distribution and exchange. In Australia the term has most commonly been used to describe the *Australian Labor Party's advocacy at various times of the nationalization of monopolies, such as *bank nationalization, and the establishment of public corporations, for example, the Commonwealth Bank (later the *Commonwealth Banking Corporation), to compete with private enterprise.

STATUTE OF WESTMINSTER *Westminster, Statute of

STEAD, CHRISTINA ELLEN (1902–83), writer, was born in New South Wales. Her father was a naturalist. She was educated at various schools, including Sydney Girls High School, business college, night schools, and Sydney Teachers College. She taught in Sydney for some time, before leaving Australia in 1928. Christina Stead spent most of her life in Britain, Europe, and the USA. She worked in various jobs, including those of Hollywood scriptwriter and instructor at New York University. She was married to William James Blake (né Blech). In 1969 she visited Australia, returned to live in Australia in 1974, and died in Sydney. She is best known for her novels, which include *Seven Poor Men of Sydney* (1934), *House of All Nations* (1938), *The Man Who Loved Children* (1940), which is perhaps her most famous work, *For Love Alone* (1944), and *Miss Herbert (The Suburban Wife)* (1976).

STEWART, ELEANOR TOWZEY (NELLIE) (1858–1931), actor and singer, was born in Woolloomooloo, Sydney, and was educated in Melbourne. Nellie Stewart, whose father was an English-born comedian and mother an Irish-born actor, first appeared on the stage in 1864. She and other members of her family toured in *Rainbow Revels* in Australia and India in the late 1870s, before she came to prominence in Australia as the principal boy in George *Coppin's pantomime *Sinbad the Sailor* in 1880. She went on to star in many operettas and pantomimes, some of which were J. C. *Williamson productions, becoming known as 'Australia's idol'. She is best known for her performances as Nell Gwynne in *Sweet Nell of Old Drury*, a role in which she appeared many times in Australia and overseas between 1902 and 1929. Nellie Stewart, who died in Sydney, published an autobiography entitled *My Life's Story* (1923).

STIRLING, SIR JAMES (1791–1865), *governor, was born in Scotland, and pursued a naval career. He sailed to Sydney with currency in 1826, explored the south-western coast of Australia in 1827, and then transferred the garrison in northern Australia from Melville Island to Raffles Bay. Stirling recommended the Swan River area as a site for British settlement, and administered the new colony of *Western Australia from its foundation in 1829 until 1832, when he visited England, and again from 1834 to 1839. (He became Governor of Western Australia in 1832, having originally been appointed Lieutenant-Governor.) He was involved in the Battle of *Pinjarra in 1834. After leaving Western Australia in early 1839, Stirling continued his naval career, eventually becoming an admiral in 1862. He died in England.

STREET (née Lillingston), JESSIE MARY GREY, LADY (1899–1970), was born at Chota Nagpur, India, the daughter of an English member of the Indian Civil Service and his wife. Jessie, whose family moved to an inherited property in northern New South Wales in 1899, was educated by governesses, at Wycombe Abbey in England from 1903 to 1906, and at the University of Sydney, from which she graduated with a BA in 1910. During a period overseas she worked for the women's suffrage movement in England. She established a model dairy on her return to Australia, helped to found the New South Wales Social Hygiene Association in 1915, married in 1916, established a Women's Sports Union at Sydney University in 1918, played an active part in organizations such as the *National Council of Women, and established a House Service Company in 1923, which trained and supplied domestic staff. Jessie Street helped to found the United Associations of Women (UAW) in 1929, which fought for legal, political, and civil rights for women, including *equal pay, and was its president for some twenty years from its foundation. Under its auspices, she founded the Australian Women's Charter in 1943 and launched the *Australian Women's Digest*, a monthly feminist magazine, in 1944. A member of the *Australian Labor Party from 1939 to 1949, she stood unsuccessfully in federal

Portrait of Nellie Stewart from a collection of photographs by Mina and May Moore. (La Trobe Collection, State Library of Victoria)

Jessie Street, the best-known Australian feminist of the 1930s and 1940s. (National Library of Australia)

elections as an endorsed candidate in 1943 and 1946 and as an independent in 1949. She organized a Russian Medical Aid and Comforts Committee, best known for its 'Sheepskins for Russia' appeal, during the *Second World War, chairing its committee from 1941 to 1945, visited the Soviet Union a number of times before and after the war, and was the first president of the Australian–Russian Society (later the Australian Soviet Society) in 1946. Jessie Street, who had been a founding member of the League of Nations Union in Australia and had attended the League of Nations in 1930 and 1938, was Australia's only female delegate at the San Francisco conference which established the *United Nations Organization in 1945. From 1947 to 1948 she represented Australia on its Status of Women Commission, which she had also helped to found. She travelled widely overseas between 1949 and 1956, during which time she attended numerous international peace conferences. After her return to Australia, she campaigned for Aboriginal rights from 1956 and was partly responsible for the holding of the 1967 referendum which gave Aborigines greater legal status. Jessie Street's father-in-law, husband, and one of her sons all were Chief Justices of the New South Wales Supreme Court. Her many publications include *Peace and Defence by Trade* (1939), *Report on Aborigines in Australia* (1957), and an autobiography entitled *Truth or Repose* (1966).

STREETON, SIR ARTHUR ERNEST (1867–1943), artist, was born in Victoria. His father was a schoolteacher. Streeton was apprenticed to a lithographer, and studied at the National Gallery School in Melbourne. He became one of the leading members of the *Heidelberg School. According to Patrick McCaughey, 'Streeton was the most various, the most unpredictable and, arguably, the most naturally gifted member of the Heidelberg School'. He contributed forty works to the controversial *Exhibition of 9 x 5 Impressions*, and later held a number of successful Australian exhibitions of his own. Streeton's early landscapes are now considered to be his best works. From 1898 he lived in England, returning to Australia for several exhibitions in 1907. During the First World War he served in the Royal Army Medical Corps and as an Australian official war artist. Streeton lived in Australia again from 1919 to 1921. He also lived in the USA for some time, before settling permanently in Australia in 1924. Streeton's paintings include *Still Glides the Stream and Shall Forever Glide* (1889), *'Fire's On', Lapstone Tunnel* (1891), and *Purple Noon's Transparent Might* (1896).

STRUTT, WILLIAM (1825–1915), artist, was born in England. He studied art in Paris, and completed many illustrations for books, before arriving in 1850 in Melbourne, where John Pascoe Fawkner became his patron. Strutt remained in Australia until 1862, except for a period in 1855–6 when he lived in New Zealand. He produced many engravings and lithographs, some of which he was employed to do for the *Illustrated Australian Magazine*; sketched and painted numerous significant events; and painted a number of portraits, including those of Fawkner, Frederick Sargood, *Coppin, and John O'Hara Burke. Strutt returned to England in 1862, and became well known for his paintings of animals. His best-known works about Australia are his history paintings, such as *Black Thursday* (1862–4) and *Bushrangers on*

*William **Strutt**'s Portrait of John Pascoe Fawkner, founder of Melbourne (1851). (National Library of Australia)*

the St. Kilda Road (c. 1887). The Library Committee of the Parliament of Victoria published Strutt's *Victoria the Golden: Scenes, Sketches and Jottings from Nature, 1850–1862*, in 1980. The Parliament's Library has held the original album since 1907.

STUART, JOHN MCDOUALL (1815–66), explorer, was born in Scotland. His father was an army captain. Stuart was educated at the Scottish Naval and Military Academy. In 1839 he arrived in South Australia, where he worked as a surveyor. Stuart was a draughtsman with Charles *Sturt's expedition to central Australia (1844–6). In 1858 he explored an area near Lake Torrens and Lake Gairdner. In 1859 he led an expedition to search for a permanent route to the north, travelling to an area beyond Lake Eyre, which he surveyed in another expedition later that year. In 1860 he made his first attempt to cross the continent from south to north; he reached the centre of Australia, and discovered nearby Central Mount Sturt (later Stuart). His second attempt, funded by the South Australian Government and conducted in competition with the *Burke and Wills expedition, took place in the first half of 1861. It also failed. His third attempt, made in 1861–2, was successful. Burke and Wills had beaten him, but died on their return trip. The *Overland Telegraph Line was built along Stuart's successful route. Stuart, almost blind from the hardships he had experienced, left Australia for Scotland in 1864, and died in London.

Stump-jump plough An agricultural implement, which behaves as its name suggests. It was significant in the development of the Mallee areas of New South Wales, South Australia and Victoria. A farmer named Mullins apparently devised the theory behind it, and R. B. and C. H. Smith put the theory into practice. The stump-jump plough was registered in 1877 and patented in 1881, and was still being used in the late twentieth century.

Sturt, Charles (1795–1869), explorer, was born in India, the son of a judge in Bengal. He was educated in England, entered the army, and served in Europe and Canada before arriving in Sydney as a captain in 1827. Sturt was largely responsible for solving the 'riddle of the rivers' in eastern Australia. On his first expedition, in 1828–9, he explored the Macquarie River, discovered the Darling River, and explored part of the Castlereagh River. On his second expedition, in 1829–30, he explored much of the Murrumbidgee River, entered the Murray River (which he named, unaware that it was the same river that Hume and Hovell had discovered further upstream, and had named the Hume) and followed it to its mouth on the southern coast (exploring part of the Darling River at its junction with the Murray along the way), then returned along the rivers. Sturt went temporarily blind as a result of hardships experienced during this expedition. He served on *Norfolk Island for some time, before returning on sick leave to England, where he retired from the army.

In 1835 he returned to New South Wales. In 1838, while overlanding cattle to Adelaide, he explored the rest of the Murray River. In 1839 he returned to Adelaide, where he expected to become Surveyor-General, but due to some confusion became Assistant Commissioner of Lands. He was later demoted to Registrar-General. On his final expedition, from 1844 to 1846, he explored part of central Australia. During this time he was appointed Colonial Treasurer. He left for England in 1847, on leave, and returned to Adelaide in 1849. He soon became Colonial Secretary, resigned in 1851, and returned permanently to England in 1853. Sturt's publications include *Two Expeditions into the Interior of Southern Australia 1828–31* (1833) and *Narrative of an Expedition into Central Australia* (1849).

'Such is life' A famous phrase, reputedly Ned *Kelly's final words before being hanged at the Melbourne Gaol on 11 November 1880. There is some disagreement as to whether he actually said this, or 'Ah well, I suppose it has come to this'. The phrase is also the title of Joseph *Furphy's well-known book.

Sudan Contingent A contingent sent from New South Wales to support the British in the Sudan. After news of General Gordon's death at Khartoum reached Australia, the New South Wales Government offered, in February 1885, to send troops. The offer was accepted. Further offers from Victoria, South Australia, and Queensland were declined. The contingent, consisting of about 750 male volunteers, accompanied by 200 horses, sailed from Sydney in the *Australasian* and the *Iberia* on 3 March 1885, and arrived at Suakin on the Red Sea on 29 March 1885. The men saw some fighting at Tamai, and three were slightly wounded, but they mostly worked on railway fatigues. Fifty formed part of a camel corps. The contingent sailed again, in the *Arab*, on 17 May 1885, arrived in Sydney on 23 June, and disbanded shortly afterwards. Nine of the Australians involved died from disease.

SUDDS AND THOMPSON AFFAIR A controversial affair which occurred in New South Wales in 1826. Joseph Sudds and Patrick Thompson, who were privates in the 57th Regiment, were convicted of theft and each was sentenced to seven years' *transportation. Governor *Darling, wishing to stamp out the practice of soldiers obtaining discharges from the *army by committing minor civil crimes, commuted their sentences to seven years' hard labour. Sudds and Thompson were paraded in iron collars, chains and leg-irons, and drummed out of their regiment. Sudds died five days later. It has been argued by some that Darling's actions were illegal. Later he was ordered to release Thompson. Darling's opponents used the affair to attack him, and *Wentworth threatened to impeach him. Parts of the press pursued the issue for political reasons. Darling responded by attempting to legislate against the press, and the *Newspaper Regulating Act 1827* was passed. Thompson was released in mid-1827 and returned to his regiment. After Darling's return to England, a select committee inquiring into his conduct in New South Wales exonerated him in 1835.

SUFFRAGE The 'right of voting in political elections'. Manhood suffrage for the lower houses was introduced in South Australia in 1856, Victoria in 1857, New South Wales in 1858, Queensland in 1859, Western Australia in 1893, and Tasmania in 1900. Adult suffrage (that is, for women as well as men) for the lower houses was introduced in South Australia in 1894, Western Australia in 1899, New South Wales in 1902, Tasmania in 1903, Queensland in 1905, and Victoria in 1908. Adult suffrage for the Commonwealth House of Representatives and Senate was introduced in 1902 (those people eligible to vote for the lower house in each state had been eligible to vote at the first federal election).

SUNSHINE HARVESTER *McKay, Hugh Victor

SUTHERLAND, DAME JOAN (1926–), soprano, was born and educated in Sydney, where she made her operatic debut in Goossen's *Judith*. She went to London in 1951, studied with Clive Carey at the Royal College of Music, and made her Covent Garden début as the First Lady in Mozart's *The Magic Flute* in 1952. She married Richard Bonynge in 1954, who encouraged her to sing coloratura and conducted most of her performances for many years. Her performance in the title role of Donizetti's *Lucia di Lammermoor*, at Covent Garden in 1959, and La Scala, Milan, in 1961, confirmed her reputation as a great soprano. She also made highly successful débuts in Italy, France, Austria, and the USA in 1960. Known as 'La Stupenda', she specialized in bel canto ('style of operatic singing concentrating on beauty of sound and vocal technique') roles, many of which were specially revived for her, and the operas of Handel. She continued to work mostly abroad, but returned to Australia to sing a number of times, the first of which was in 1965, when she brought her own opera company for a season organized by the company founded by J. C. *Williamson. Created a dame in 1979, she retired in 1990, her final performance being as Queen Marguerite de Valois in Meyerbeer's *Les Huguenots* at the *Sydney Opera House, a venue at which she had first sung in 1973.

SUTHERLAND, MARGARET (1897–1984), composer, was born in Adelaide, but spent most of her life in Melbourne. Her father was a journalist. She studied at the Albert Street Conservatorium of Music, Melbourne (later the Melba Memorial Conservatorium), and went to Europe in 1923, spending time in London (where she studied composition under Arnold Bax) and Vienna, before returning to Australia in 1925. Margaret Sutherland composed some twenty orchestral works, an opera entitled *The Young Kabbarli* (1965), based upon experiences of Daisy *Bates, more than twenty chamber works, two ballets, and a number of piano, choral, vocal, and incidental works. She also organized midday chamber music concerts for the Red Cross during the Second World War and played a leading role in the Combined Arts Centre Movement, the object of which was to have an arts centre established on land in St Kilda Road (the site of the present Victorian Arts Centre), during the 1940s and 1950s. She was a founding member of the Council of Education, Music and the Arts in the 1940s; a council member of the National Gallery of Victoria during the 1950s and 1960s; and a member of both the Australian Unesco Music Committee and the Advisory Board of the Australian Music Fund during the 1960s.

SWAGMEN were itinerant labourers. Their name came from the swags, or bundles of personal belongings, which they carried. They tramped around country areas, seeking employment on stations in return for supplies. Many travelled alone. They were most common at the turn of this century and during the *Great Depression of the 1930s. Swagmen have become part of Australian folklore.

SWAN RIVER *Western Australia

SWIMMING has been one of Australia's most popular sports for many years. The first swimming championship to be held in the Australian colonies took place in Sydney in 1846, and the Amateur Swimming Union of Australia was founded in 1909. Since the 1890s Australian swimmers have competed in international competitions with considerable success, particularly in the *Olympic Games held in Melbourne in 1956 (at which Australia won eight gold medals for swimming). Notable Australian swimmers have included members of the Cavill family, one of whom was Richard Cavill (1884–1938), said to have been the first person to use the stroke that became known as the Australian crawl in a competition (in 1899); Frederick Lane (1880–1969); Sir Frank Beaurepaire (1891–1956); Annette Kellermann (1886–1975); Sarah (Fanny) Durack (1889–1956); Andrew ('Boy') Charlton (1907–75); Lorraine Crapp (1939–); Dawn Fraser (1937–); Murray Rose (1939–); John Konrads (1942–); Ilsa Konrads (1944–); and Shane Gould (1957–). Non-competitive swimming also has been, and continues to be, very popular in Australia. Surfing, competitive and non-competitive, has also

Swagmen have been the subject of many literary and artistic works, such as Lionel Lindsay's Swagman Bending to Lift a Billy *(1936). (National Gallery of Victoria)*

become increasingly popular among Australians, especially since the 1960s.

'SWING' RIOTERS English agricultural labourers who took part in the 'Swing' Riots of 1830, agrarian riots in wheat-growing areas such as Hampshire, Wiltshire, and Berkshire. They smashed machines, burnt ricks, rioted over wages, and sent threatening letters to farmers in protest over poor conditions and the introduction of new threshing machines, which they saw as a threat. They took their name from that of their mythical

'Swing' Rioters

Swimming scenes at St Kilda and Elwood, in Melbourne, from the Australasian, 1 December 1923. (La Trobe Collection, State Library of Victoria)

leader, 'Captain Swing', whose signature often appeared on the threatening letters. They are known also as the machine-breakers. After the riots, 1976 people appeared before ninety courts. Nineteen were hanged; 476, two of whom were women, were transported to Australia—332 to Van Diemen's Land, the remaining 144 to New South Wales. Almost all arrived in the colonies in

SYDNEY OPERA HOUSE

1831 and were assigned. Most received free pardons between 1836 and 1838. Very few appear to have returned to England. The 'Swing' Rioters formed the largest single group of protesters to be transported to Australia.

SYDNEY, the capital city of *New South Wales, is Australia's oldest city. *Phillip, having rejected *Botany Bay as the site for Britain's new penal colony, explored to the north and chose a site on Port Jackson. He named it Sydney Cove, after the British Secretary of State for the Colonies. Settlers from the *First Fleet landed there on 26 January 1788. Sydney was proclaimed a city in 1842. Sydney's population was slightly more than 1000 in 1788, 224 939 in 1881, and 3 596 000 in 1988.

SYDNEY HARBOUR BRIDGE A toll bridge, which crosses Sydney Harbour from Dawes Point to Milson's Point. John Job Crew Bradfield was largely responsible for its design. Its construction began in 1923. Jack *Lang, the New South Wales Premier, officially opened it on 19 March 1932, although Francis Edward de Groot, a member of the *New Guard, slashed the ribbon before him (and was consequently arrested).

SYDNEY MORNING HERALD *Fairfax, John

SYDNEY OPERA HOUSE A striking white building, in appearance like a series of wind-filled sails, situated at Bennelong Point on the edge of Sydney Harbour. Joern Utzon, a Danish architect, was largely responsible for its design. In 1957 he won a competition to design it run by the New South Wales Government, but resigned amid controversy in 1966. Construction, mostly funded by lotteries, began in 1959 and ended in 1973. The building houses a concert hall, drama and opera theatres, restaurants, and other facilities.

City Improvements in Sydney, *from the* Australasian Sketcher, *31 July 1880, showing old buildings ordered for demolition in Sydney.* (La Trobe Collection, State Library of Victoria)

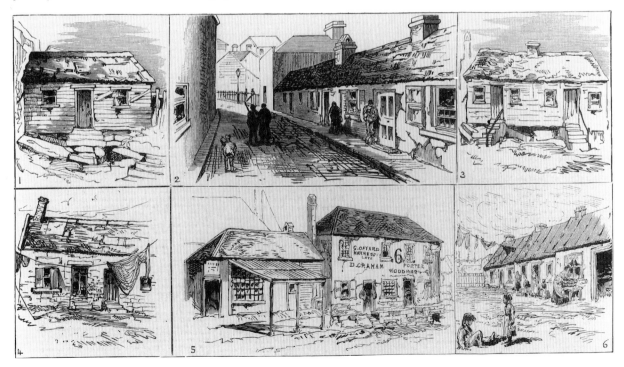

Roland Wakelin's The Bridge Under Construction (c. 1928–9), showing part of the **Sydney Harbour Bridge** before it was completed. (National Gallery of Victoria)

SYME, DAVID (1827–1908), newspaper proprietor, was born in Scotland and educated by his father, who was a schoolmaster. He also travelled and studied in Germany. Syme went to the Californian gold-fields in 1851, then was a *digger on the Victorian gold-fields from 1852 to 1855, before becoming a road contractor. His brother, Ebenezer Syme (1826–60), bought the Melbourne *Age* in June 1856. David Syme, who became Ebenezer's partner in September 1856, published and edited the *Age* from 1859 until his death. His policies included the opening of land for *selectors, the *protection of local industries, the introduction of democratic reforms, support for *free, compulsory, and secular education, and reforms in working conditions. He wielded considerable political influence, especially during the 1890s. The newspaper's circulation increased from about 2000 in 1860 to about 120 000 in 1899. After his death, his sons continued the newspaper as a trust, which was superseded by a public company in 1948. Syme's publications include *Outlines of an Industrial Science*

(1876), *Representative Government in England* (1881), *On the Modification of Organisms* (1890), *The Soul: A Study and an Argument* (1903), and various journal articles on political economy.

SYSTEMATIC COLONIZATION A socio-economic theory expounded by Edward Gibbon *Wakefield and others, which influenced British colonial policies, particularly during the 1830s and 1840s. The theory proposed a balance between land, labour, and capital. Colonial lands should be sold at a so-called 'sufficient price', which would delay, but not stop, labourers from becoming land-owners. Capital raised could be used to finance emigration from the mother country. Colonies eventually would become self-governing. Theoretically, the system would be self-regulating, and of mutual benefit to mother country and colonies. Some principles of this theory were applied to the White settlement of Australia, in particular to that of *South Australia. Wakefield was critical of the way in which South Australia was settled, and dissociated himself from it.

Caricature of David Syme by Phil May. (David Syme & Co. Ltd)

T

TAA *Trans Australia Airlines

TAILORESSES' STRIKE A strike in 1882–3 said to have been the first organized women's strike in Australian history. Tailoresses employed in Melbourne factories went on strike over alleged sweating. The main periods of the strike were in December 1882 and February 1883. The Trades Hall Council, the *Age*, and many members of the public supported the strikers. Estimates of the numbers of striking tailoresses range from about four hundred to two thousand. Most of the employers involved agreed to the demands of the Tailoresses' Union in February 1883. Some writers claim that the Tailoresses' Union was formed shortly after the strike began, others that it was already in existence.

TASMAN BRIDGE A bridge across the Derwent River estuary, opened in 1964, linking the two sides of *Hobart. Part of the bridge (two pylons and three spans) collapsed on 5 January 1975 when an Australian National Line bulk carrier, the *Lake Illawarra*, hit it. Seven crew members, and five people trapped in cars which plunged into the water, were killed. On 30 April 1975 a marine court of inquiry found the *Lake Illawarra*'s master, Captain Boleshaw, solely responsible for the disaster. The bridge reopened on 8 October 1977.

TASMAN, ABEL JANSZOON (1603?–59), Dutch mariner, was born near Groningen, in the Netherlands. He joined the Dutch East India Company in 1632, and participated in various military, trading, and exploratory voyages before being appointed to command a voyage to explore the southern and eastern seas in 1642. On this voyage, he discovered *Van Diemen's Land, which he claimed for the Dutch and named after the Governor-General of the Dutch East Indies (part of present-day Indonesia); he also discovered *New Zealand, the Tonga Islands, and some of the Fiji Islands; he returned to Batavia (present-day Jakarta) along the northern coast of *New Guinea. In 1644 Tasman led another voyage to the area. He confirmed that New Guinea and *New Holland (present-day Australia) were not joined, but that Carpentaria (part of present-day Queensland), and De Witt's Land (part of present-day Western Australia) each belonged to the same land mass. He charted the northern and north-western coast from the Gulf of Carpentaria to present-day North West Cape. After returning to Batavia, he was confirmed as *commandeur* and was appointed to the Council of Justice. He led a diplomatic mission to Siam (present-day Thailand) in 1647. He then led a fleet of eight vessels against the Spanish in 1648, but the following year was removed from office for maltreating a sailor on that voyage. He was reinstated in 1651, but retired soon afterwards, and became a merchant in Batavia, where he died. Van Diemen's Land was renamed *Tasmania after him in 1855.

TASMANIA, a state of the *Commonwealth of Australia since the latter's creation in 1901, consists of a large island, separated from mainland Australia by Bass Strait, and a number of smaller nearby islands. In its early years of White settlement, which began in 1803, it was known as *Van Diemen's Land. In 1855 the colony was renamed Tasmania, after *Tasman, and granted *responsible government. The first responsible ministry took office the following year. The development of Tasmania from the 1930s was greatly

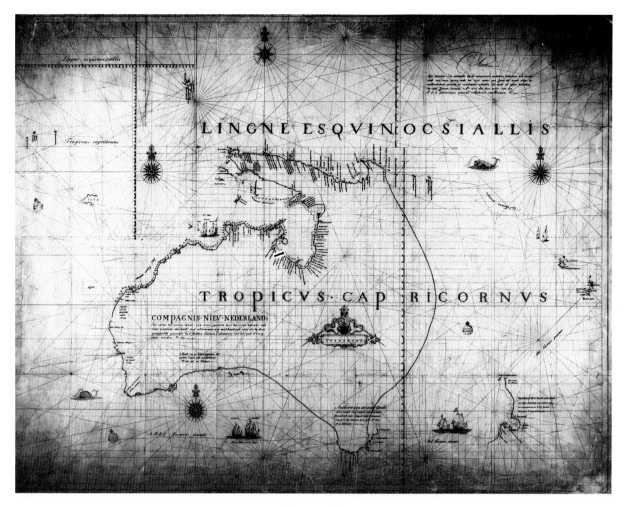

Map of Australia, New Guinea, and New Zealand showing the results of Tasman's voyages of 1642–3 and 1644. (Mitchell Library, State Library of New South Wales)

assisted by the *Commonwealth Grants Commission.

TATTERSALL'S SWEEPSTAKES   Popular Australian lotteries based upon horse-races. In 1878 George Adams (1839–1904) took over O'Brien's Hotel (which was also the headquarters of the Tattersall's Club, modelled upon Tattersall's in London), in Pitt Street, Sydney. He continued to operate the private sweepstakes which had been held at the hotel for some time, and first held a public sweepstake on the Sydney Cup in 1881, selling two thousand tickets at one pound each. After the New South Wales Government made sweepstakes illegal, Adams moved Tattersall's to Brisbane in 1893, but similar problems arose there, causing him to move it to Hobart in 1895. At the request of the Tasmanian Government, he conducted two lotteries, in 1895 and 1896, to dispose of the assets of the failed Bank of Van Diemen's Land. The first Tattersall's sweepstake on a horse-race in Hobart was held on the Anniversary Cup on *Australia Day 1896. The following year, Tattersall's became the official state lottery of Tasmania. The business continued after Adams's death in 1904. Regardless of federal postal bans from 1902 to 1930 on mail addressed to Tattersall's, having 'a ticket in Tatt's'

This Tasmanian Tourist Association's advertisement, Holidays in Southern Tasmania, *from the Tasmanian Mail, 22 December 1906, presents an intriguing catalogue of the delights of **Tasmania** in the early years of the twentieth century. (Archives Office of Tasmania)*

became an Australian institution. In 1954, the organization's headquarters were moved to Melbourne. Tattersall's introduced Tattslotto, a numbers game, in 1972 and various other gambling games in later years.

TEBBUTT, JOHN (1834–1916), astronomer, was born at Windsor, New South Wales, the son of a farmer. Influenced by Edward Quaife, one of his teachers at the local Church of England parish school, Tebbutt became interested in

'Temper, democratic'

Ticket for one of the lotteries, conducted by George Adams of Tattersall's sweepstakes, to dispose of the Bank of Van Diemen's Land. (Tattersall Sweep Consultation)

astronomy at an early age. In May 1861 he discovered the comet which was named after him. He declined the position of Government Astronomer of New South Wales in 1862, but built his own observatory at Windsor in 1863 (and another one near by in 1879), and continued his observations, which included ones on the transit of Venus in 1874 and the first sighting of the great comet of 1881. Tebbutt, who was elected a Fellow of the Royal Astronomical Society (London) in 1873, was the first president of the New South Wales branch of the British Astronomical Association in 1895. He died at Windsor. His publications include *Meteorological Observations made at the Private Observatory of John Tebbutt, Jnr.* (1868), *History and Description of Mr. Tebbutt's Observatory* (1887), *Astronomical Memoirs* (1908), and some 370 journal articles.

'Temper, democratic; bias, offensively Australian' A now-famous phrase, first used by *Furphy to describe his novel, *Such is Life: Being Certain Extracts from the Life of Tom Collins* (1903). The complete sentence read: 'I have just finished writing a full-sized novel: title, *Such is Life*; scene, Riverina and northern Vic; temper, democratic; bias, offensively Australian.' Furphy wrote this in a letter that accompanied his manuscript when he sent it to the Sydney *Bulletin* for advice in 1897.

Tench, Watkin (1758?–1883), soldier and writer, was born in England. He entered the marine corps in 1776, and served in the American War of Independence, during which he was a prisoner-of-war for some months. He sailed to New South Wales with the *First Fleet, as one of the two captain-lieutenants of the detachment of marines. He remained in the colony until 1792, when he sailed to England with the marines. Tench was again a prisoner-of-war for some time during the war with France. He retired, with the rank of lieutenant-general, in 1821. He died in England. Tench is best known in Australia for two books, both based upon his journal: *A Narrative of the Expedition to Botany Bay, with an Account of New South Wales, its Productions, Inhabitants &c* (1789) and *A Complete Account of the Settlement at Port Jackson, in New South*

Wales, including an Accurate Description of the Situation of the Colony, and of its Natural Productions, taken on the Spot (1793).

TENNIS has been for many years one of Australia's most popular sports, for both spectators and participants, with access to tennis-courts being widely available. It was first introduced to the Australian colonies during the 1870s. The Lawn Tennis Association of Australia (originally Australasia) was founded in 1904. The Australian championships, first played in 1905, form part of the 'grand slam', the four main international tennis championships for players competing as individuals (the others being held at Wimbledon, in France, and in the USA). Many Australians have been successful in these championships over the years, particularly during the 1950s and 1960s. Australia (in partnership with New Zealand until 1923) first competed in the Davis Cup (the International Lawn Tennis Challenge Trophy) competition, the main men's international tennis team competition, in 1905, and won twenty-four times (six of those with New Zealand) up to and including 1977. Australia also won the Federation Cup, the sport's main women's international team competition, seven times between 1963 (when that competition began) and 1978. Australia's many notable tennis players have included Sir Norman Brookes (1877–1968), Ken Rosewall (1934–), Lew Hoad (1934–), Roy Emerson (1936–), Rod Laver (1938–), Margaret Court (née Smith) (1942–), John Newcombe (1944–), and Evonne Cawley (née Goolagong) (1951–).

TENT EMBASSY *Land Rights

TENTERFIELD ORATION A now-famous speech given by Sir Henry *Parkes, the Premier of New South Wales, at Tenterfield, New South Wales, on 24 October 1889. He argued for the federation of the Australian colonies, largely for defence reasons.

TERRA NULLIUS A Latin term meaning land belonging to no one, or unoccupied land. At the time of White discovery and settlement of *New South Wales, the British apparently considered that the area was *terra nullius*, because the *Aborigines did not occupy the land in a European sense. The British believed that if the land was *terra nullius* (according to their definition), then the first Europeans to discover it, and to occupy it, were entitled to it.

THEODORE, EDWARD GRANVILLE (1884–1950), politician and business man, was born at Port Adelaide, South Australia. His father was a tugboat operator. Theodore left school at twelve and worked in various local jobs, then in mining and other jobs in several states. His union involvement included founding the Amalgamated Workers' Association in Queensland in 1907, and being president of the *Australian Workers Union, also in Queensland, in 1913. Theodore was a Labor Member of the Queensland Legislative Assembly from 1909 to 1925. He held various portfolios, and was Premier from 1919 to 1925, when he resigned in order to enter federal politics. Theodore, who had become known as 'Red Ted', was a Labor Member of the House of Representatives from 1927 until he was defeated in the elections held in December 1931. He was Deputy Leader of the Labor Party, and of the Opposition in 1929, and Treasurer in the *Scullin Government. He resigned from the ministry in July 1930, following allegations by a Queensland royal commission that his conduct relating to the sale of the Mungana mines, while Premier of Queensland, had been fraudulent and dishonest. His reinstatement in January 1931 was an immediate factor in a major split in the *Australian Labor Party. Later in 1931 a court exonerated him over his actions in the Mungana affair. As Treasurer, Theodore advocated a mildly inflationary approach to deal with the *Great Depression. Following his electoral defeat he pursued business interests. During the Second World War he was Director-General of Allied Works from 1942 to 1944. He died in Sydney.

THIRD FLEET Eleven ships, carrying convicts and supplies, which sailed from England and Ireland to New South Wales in 1791, after the *First Fleet and *Second Fleet. The ships of the Third Fleet were the *Active*, the *Admiral Barrington*, the *Albermarle*, the *Atlantic*, the *Britannia*, the *Gorgon* (a supply ship), the *Mary Ann*, the *Matilda*, the *Salamander* and the *William and Ann*. Some 1865 convicts began the voyage, but about 185 of them died before reaching New South Wales.

'THIRTY-SIX FACELESS MEN' A derogatory phrase that was used to describe the thirty-six member *Australian Labor Party federal conference, and its supposed relationship with the Party's federal parliamentary leadership. There is some debate as to who first used the phrase; it has usually been attributed to the journalist Alan Reid. A special federal conference met in 1963 to decide party policy on a proposed United States naval communications base at North West Cape. At that time the parliamentary leader (*Calwell) and the deputy leader (*Whitlam) did not have the right to attend such conferences. Press photographs showed Calwell and Whitlam having to wait outside while the conference decided party policy. *Menzies used the phrase during the federal election campaign later that year, claiming that the Australian Labor Party was subject to anonymous external control. The incident led to the Party's parliamentary leader and deputy leader gaining the right to attend and to speak, although not to vote, at federal conferences.

THORNE KIDNAPPING The first known case of kidnapping for ransom in Australia. Graeme Thorne, an eight-year-old boy whose parents had recently won an Opera House Lottery, was kidnapped from the *Sydney suburb of Bondi on 7 July 1960. A ransom of £25 000 was demanded. After Thorne's body was found in Seaforth on 16 August of the same year, Stephen Bradley (né Istvan Baranyay) was arrested, found guilty of killing Thorne, and sentenced to life imprisonment. He died while in gaol. The case attracted considerable publicity.

'THREE ELEVENS' A now-famous phrase used by *Deakin in a speech given in Melbourne on 1 February 1904. He likened the situation in the Commonwealth Parliament to a cricket game with 'three elevens' competing instead of two. After the general election held on 16 December 1903, the seats were distributed almost equally among the three parties. In the House of Representatives, the Protectionists held twenty-five seats, the Free Traders twenty-four, Labor twenty-five, and there was one

Independent; in the Senate, Labor held fourteen, the Free Traders thirteen, and the Protectionists nine. Deakin's full sentence ran as follows: 'What kind of a game of cricket, compared with the present game, could they play if they had three elevens instead of two, with one playing sometimes with one side, sometimes with the other, and sometimes for itself?'

TICHBORNE CASE A sensational nineteenth-century English legal case. Roger Tichborne, the heir to a baronetcy and fortune, disappeared in a shipwreck off the South American coast in 1854. His mother later advertised in search of him. One Thomas Castro, a butcher's assistant from Wagga in New South Wales, responded to the advertisement in 1865, claiming to be Tichborne. The Tichborne Claimant, as he was called, went to England in 1866. He convinced Tichborne's mother, but not the rest of the family, that he was Tichborne. A civil action to establish his claims failed in 1872. He then was charged with perjury, found guilty, and sentenced to fourteen years' imprisonment. It appears that he was a butcher's son named Arthur Orton, who was born in Wapping, London, in 1834. He went to South America in 1849, returned to England in 1851, and then sailed for Australia in 1852. He had various jobs in Australia, before becoming a butcher's assistant (under the name of Castro) in Wagga Wagga in 1864. After his release

*A convict's **ticket of leave passport**. (National Library of Australia)*

from jail in 1884, he became a music-hall actor. In 1895 he 'confessed' to being an imposter, but later withdrew his 'confession'. He died in London in 1898. The Tichborne case attracted much press and public interest, in both England and Australia.

TICKET OF LEAVE A certificate that could be granted to a convict during the *transportation period. It allowed the convict to be excused from compulsory labour, to choose his or her own employer, and to work for wages. There were some restrictions, and the ticket of leave could be withdrawn. It was usually granted for good conduct, but sometimes for other reasons, such as economy.

TOLPUDDLE MARTYRS (also known as the Dorchester Labourers) The popular name given to six agricultural labourers from the village of Tolpuddle, in England, who were transported to Australia in 1834. They were James Brine, James Hammett, James Loveless, George Loveless (the group's leader), John Standfield, and Thomas Standfield. Their ostensible crime was 'administering unlawful oaths'; their actual 'crime' was organizing a trade union (the Friendly Society of Agricultural Labourers), although it was not illegal to do so. The six were tried at the Dorchester Assizes, were found guilty, and each was sentenced to seven years' transportation. All were transported to New South Wales, except for George Loveless, who was transported to Van Diemen's Land. All were assigned to settlers. After much protest in England, they were granted free pardons and free passages in 1836, and all returned to England between 1837 and 1839. Hammett alone returned to Tolpuddle; the others farmed in Essex for about two years, before migrating to Canada. George Loveless described the experiences of the Tolpuddle Martyrs in *The Victims of Whiggery* (1837).

TORRENS SYSTEM A system for registration of ownership of land. It takes its name from Robert Richard Torrens, formerly Collector of Customs and Registrar-General of Deeds, who introduced a Bill incorporating it in the South Australian Parliament in 1857 during his brief term as Premier. The Bill was passed in 1858. The system simplified the transfer of land through the establishment of a government register guaranteeing land ownership. Under the old system, the onus to prove ownership had rested with individuals. It was adopted in Queensland, New South Wales, Victoria and Tasmania during the 1860s, and in Western Australia in 1874. It was also adopted in other countries.

TOWNS, ROBERT (1794–1873), merchant, was born in England, and went to sea when young. From 1827 he visited Australia many times in charge of vessels, sometimes his own. He settled in Sydney in 1843, where he already held investments and property, and became a successful merchant and ship-owner, trading in the Pacific and elsewhere. He also acquired much land, particularly in Queensland in the 1860s. Towns introduced *blackbirding to Australia in 1863, when he began importing Pacific Islanders to work on his cotton plantations near Brisbane; the practice was to play a significant role in the development of the Australian sugar industry. In 1864 J. M. Black, an agent acting for Towns, established a settlement at Cleveland Bay, which formed the basis of the town (later city) and port of *Townsville. Towns was a director of the *Bank of New South Wales (1850–5, 1861–7); was involved in its reorganization in 1851; and was its president twice (1853–5, 1866–7). He was a Member of the New South Wales Legislative Council (1856–61, 1863–73), and was president of the Sydney Chamber of Commerce several times. He wrote *South Sea Island Immigration for Cotton Culture* (1863).

TOWNSVILLE is a city and port in northern Queensland, located on Cleveland Bay. J. M. Black, acting for Robert *Towns, established a settlement (which he originally called Castletown) on the site in 1864. Townsville was gazetted a town and port in 1865, was declared a municipality in 1866, and became a city in 1903. Queensland gold discoveries in the 1870s, particularly at Charters Towers, greatly assisted Townsville's development. An artificial harbour was built before the end of the nineteenth century. Australian and American troops were stationed there during the Second World War. Townsville is connected by rail with Brisbane (1340 kilometres to the south-east), Cairns, Charters Towers, *Mount Isa, and Cloncurry, and the port serves mining, pastoral, sugar and other industries.

TRADE DIVERSION POLICY A controversial trading policy introduced by the *Lyons *United Australia Party–Country Party Government in 1936. In theory trade was to be diverted away from bad customers (that is, those that sold much more to Australia than they bought) towards those that were good customers.

*An unknown photographer recorded the opening of the cathedral at **Townsville** in 1871. (National Gallery of Victoria)*

Trans Australia Airlines

Poster by James Northfield, Yesterday Cobb & Co.—Today TAA, *for Trans Australia Airlines.* (La Trobe Collection, State Library of Victoria)

Australian industry also was to be protected, using bounties, duties, and import licences. In practice, the main targets were Japan (a good customer) and the USA (a bad customer); trade in textiles was to be diverted from Japan towards Britain, and trade in motor vehicle bodies was to be diverted from the USA towards locally based, British-owned, manufacturers. Britain (a good customer) was the main beneficiary. This policy was dropped within about two years. Japanese and American exports to Australia quickly returned to pre-1936 levels, while their imports from Australia remained at much lower levels.

Trans Australia Airlines (TAA) For many years one of Australia's two main domestic airlines, the other being *Ansett Airlines of Australia. The *Chifley Labor government attempted to nationalize civil aviation through the establishment of the Australian National Airlines Commission in 1945. The attempt failed when the *High Court found it to be unconstitutional. However, Trans Australia Airlines, run by the Australian National Airlines Commission, was founded in 1946, with a daily return service between Melbourne and Sydney. The extent of the services was soon increased and TAA became very successful. In 1952 the *Menzies Liberal–Country Party Government introduced a new policy regarding Australia's domestic airlines, which later became known as the 'two-airline policy': the two airlines were TAA and its chief competitor, the privately owned, and ailing, Australian National Airways (ANA). When Ansett Transport Industries took over ANA in 1957, the 'two-airline policy' was applied to

Ansett–ANA (later known as Ansett Airlines of Australia) and TAA. In 1986 TAA's name was changed to Australian Airlines. The 'two-airline policy' was ended in 1990 with the deregulation of the domestic airline industry.

TRANSPORTATION, as a form of punishment, was used in Britain from the seventeenth century until 1868. Most of the convicts were sent to Britain's American and Australian colonies. Transportation to America began in the mid-1660s and ceased in 1776, because of the American War of Independence. About 161 700 convicts (of whom nearly 25 000 were women) arrived in the Australian colonies between 1788 and 1868. Three-quarters of them came from England, Scotland, and Wales; a little under a quarter came from Ireland; and a few hundred came from India, Canada, and other British colonies. The most common crime for which they were transported was theft. There were about 1000 political prisoners, including the *Scottish Martyrs, the *Tolpuddle Martyrs, some of the *Canadian Rebels, and a number of *Chartists. About half of the convicts were sent to New South Wales (including *Moreton Bay and the *Port Phillip District) (1788–1840), and a further 3000 *exiles were sent there during the 1840s. *Van Diemen's Land (1803–53) and Western Australia (1850–68) also received convicts. In the early years of White settlement in New South Wales and Van Diemen's Land, convicts were used mainly on public works. Assignment (allotting convicts as unpaid servants to colonists) was used extensively, especially during the 1820s and 1830s. Various forms of probation were used, especially in Van Diemen's Land during the 1840s and early 1850s. Additional punishment of convicts included corporal punishment, solitary confinement, hard labour, confinement to *female factories for women, transportation to *penal settlements, and capital punishment.

TREATY OF VERSAILLES *Versailles, Treaty of

Poster opposing **transportation**, *1850. (Public Record Office, Kew, England)*

'TRUE PATRIOTS ALL' A now-famous phrase, which was used to describe convicts in the early years of *transportation. It comes from the following stanza of a prologue written by 'a Gentleman of Leicester' (now believed to have been Henry Carter, who was *not* a convict):

> From distant climes o'er widespread seas we come,
> Though not with much éclat or beat of drum,
> True patriots all; for, be it understood,
> We left our country for our country's good;
> No private views disgrac'd our generous zeal,
> What urg'd our travels was our country's weal;
> And none will doubt but that our emigration
> Has prov'd most useful to the British nation.

For many years the prologue was attributed to George Barrington, a renowned pickpocket, who had been transported to New South Wales for stealing a gold watch and chain. He was reputed to have spoken the words of the prologue at the opening of the first Australian theatre in 1796.

TRUGERNANNER (often spelt Truganini, sometimes Trugernanna, Trugernini, Truggernana) (1812?–76), Tasmanian *Aborigine, was born on the western side of the D'Entrecasteaux Channel, *Van Diemen's Land. Her father Mangerner (or Mangana) was an Aborigine from Bruny Island. She met George *Robinson at Bruny Island in 1829, and assisted him when he travelled around Van Diemen's Land rounding up Aborigines during the 1830s. Trugernanner moved to the Flinders Island settlement in 1835, where she remained, except for one trip back to the mainland in 1836, until she went to the *Port Phillip District with Robinson in 1839. In 1841 she was one of five Aborigines charged with the murder of two White *whalers near Portland Bay. Two of the Aborigines were hanged, and Trugernanner was returned to Flinders Island in 1842. The settlement was moved to Oyster Cove, on the east coast of Van Diemen's Land, in 1847. After the other Aborigines had died, Trugernanner left Oyster Cove in 1874 and went to live with a White family in Hobart, where she died. Her skeleton was displayed in the Tasmanian Museum, Hobart, between 1904 and 1951. Her bones were cremated, and her ashes spread over the D'Entrecasteaux Channel, in 1976. Trugernanner has often, but inaccurately, been described as the last Tasmanian Aborigine.

TURNER (née Burwell), ETHEL SIBYL (1872–1958), writer, was born in England. Her father, a merchant, died when she was two years old. She and her sister took their stepfather's name after their mother's subsequent remarriage. After the stepfather's death, the family left England for Australia in 1879. Ethel attended the Sydney Girls' High School, before she and her sister Lilian founded and ran the *Parthenon*, a monthly journal, between 1889 and 1892. Ethel Turner then contributed articles for children to the *Illustrated Sydney News*, and later to the *Town and Country Journal*. She wrote many books for children, the best-known of which is *Seven Little Australians* (1894), now a classic. It has sold widely, has been translated into a number of languages, and

*Thomas Bock painted this portrait of **Trugernanner** (whose name is spelt in various ways), Truggernana, Native of the Southern Part of Van Diemen's Land, for George Augustus Robinson. It appears that Bock painted it in about October 1831, while he was still a convict. (Tasmanian Museum and Art Gallery)*

has been adapted for film and television. Ethel Turner (whose married name was Curlewis) died in Sydney.

'TWO-AIRLINE POLICY' *Trans Australia Airlines

TYSON, JAMES (1819–98), pastoralist, was born in New South Wales. His father was a free settler, his mother an ex-convict. Tyson worked in various pastoral jobs before he and his brother William established a successful slaughter-yard and butcher's shop at Bendigo, during the *gold-rushes. After selling the business in 1855, Tyson began to acquire stations in New South Wales and Queensland. By 1898 he held 2 156 680 hectares, of which 142 585 hectares were freehold. He was a Member of the Queensland Legislative Council from 1893 to 1898, and a magistrate in New South Wales and Queensland.

u

UNAIPON, DAVID (1872–1967), inventor, preacher, and writer, was born at the Point McLeay mission, near Tailem Bend in South Australia, the son of an evangelist and his wife. He was educated at the mission school, and worked in various jobs at Point McLeay and in Adelaide in the 1880s and 1890s. For the next fifty years, he travelled around south-eastern Australia, working for the Aboriginals' Friends' Association, giving sermons, and speaking on Aboriginal affairs. He appeared before royal commissions relating to the latter in 1913 and 1926, and assisted the Bleakley inquiry into Aboriginal welfare in 1928–9. In 1926 he argued for the creation of a separate Aboriginal state. His inventions included a shearing hand-piece and a centrifugal motor. The problem of perpetual motion intrigued him for most of his life. Unaipon, who wrote stories and versions of Aboriginal legends, is said to have been the first Aboriginal writer to have had a book published. His publications include *Hungerrda* (1927), *Kini Ger: The Native Cat* (1928), and *Native Legends* (1929). He was awarded a Coronation Medal in 1953.

UAP *United Australia Party

UNIFORM TAXATION Popular name given to the uniform system of income taxation, of individuals and companies, that has operated throughout Australia since 1942. Under this system, the Commonwealth has an effective monopoly in the area of income taxation. The various colonies (states from 1901 onwards) introduced income taxation between 1884 and 1902. The Commonwealth did so in 1915, during the First World War. Both Commonwealth and states continued to levy income tax until 1942, different states having different rates. The *Curtin Labor Government introduced uniform taxation as a wartime measure in 1942. The original legislation provided for the transfer of personnel, equipment, and records from the state income-tax departments to the Commonwealth; the imposition of Commonwealth income tax at a rate high enough to cover the sum formerly raised by both Commonwealth and states, making it impractical for the states to continue to levy income tax; priority to be given to payment of Commonwealth income tax; and the reimbursement of lost revenue to the states, dependent upon their not levying income tax. South Australia, Victoria, Queensland, and Western Australia challenged the system's validity before the *High Court in 1942, which found it to be valid, and that only the transfer of departments was dependent upon wartime defence powers. Victoria and New South Wales also challenged the system before the High Court in 1957. The court again found it to be valid, except for the priority provision. There have been periodic suggestions that the states should levy their own income tax again, but the Commonwealth monopoly has continued. The introduction of uniform taxation greatly increased the Commonwealth's power, at the expense of the states.

UNITED AUSTRALIA PARTY (UAP) A political party formed in 1931, during the *Great Depression, when the *Nationalist Party and a group of former Labor Party members, including J. A. *Lyons, merged. The major extra-parliamentary organization behind it was the All for Australia League. The United Australia Party governed federally from 1932 to 1941; it governed alone from 1932 to 1934, and from 1939 to 1940,

UNITED NATIONS ORGANIZATION

*Agricultural science students at the University of Melbourne, one of Australia's oldest **universities**, photographed during the First World War.*

and in coalition with the Country Party for the remainder of the period. It also governed at state level, in coalition with the Country Party, for periods in New South Wales and Victoria. The federal parliamentary leaders of the United Australia Party were Lyons (1931–9), *Menzies (1939–41, 1943–5), and *Hughes (1941–3). The United Australia Party was dissolved and absorbed into the newly formed *Liberal Party of Australia in 1945.

UNITED NATIONS ORGANIZATION (UN) An international organization established in 1945, shortly before the conclusion of the Second World War. Its aims, to solve international disputes peacefully and to foster international co-operation, are similar to those of the *League of Nations, which it superseded. Australia was one of the UN's fifty-one founding members. Since then more than one hundred other countries have joined. The UN is made up of the General Assembly, the Security Council, the Economic and Social Council, the International Court of Justice, the Secretariat, the International Monetary Fund, and various related agencies. Its headquarters are in New York.

UNIVERSITIES were first set up in Australia at Sydney (1850) and Melbourne (1853), followed by Adelaide (1874), Tasmania (1890), Queensland (1909), and Western Australia (1911). The numbers of universities increased greatly after the Second World War. The Aus-

tralian National University was established, originally as a postgraduate institution, in 1946, and subsumed the Canberra University College (formerly affiliated with the University of Melbourne) in 1960. The University of New South Wales (known as the New South Wales University of Technology until 1958) was established in 1949, Monash University (Melbourne) in 1958, Macquarie University (Sydney) in 1963, La Trobe University (Melbourne) in 1964, Murdoch University (Perth) in 1973, Griffith University (Brisbane) in 1971, and Deakin University (Geelong, Victoria) in 1976. Branches of other universities that became autonomous include the University of New England (Armidale, New South Wales) in 1954; Newcastle University in 1965; Flinders University (Adelaide) in 1966; the James Cook University of North Queensland (Townsville) in 1970; and the University of Wollongong in 1975. Total numbers of students enrolled in Australian universities increased from 14 236 in 1939 to 200 000 in the mid-1980s. From the late 1980s onwards, a number of Australian universities began to amalgamate with other higher education institutions.

V

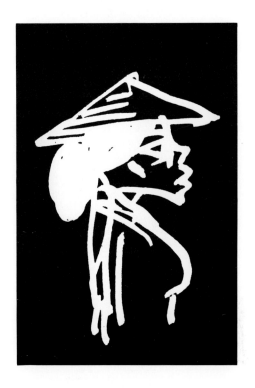

VAN DIEMEN'S LAND The name previously given to present-day Tasmania. *Tasman, the Dutch mariner, was the first White person to discover the island, in 1642, and he named it Anthony Van Diemen's Land, after the Governor-General of the Dutch East India Company. Other Europeans, including Marion du Fresne (1772), James *Cook (1777), Joseph-Antoine Raymond Bruny d'Entrecasteaux (1792–3), *Bass and *Flinders (1798–9), and *Baudin (1802), followed. British convict settlements were established in the south and north of the island. John Bowen, sent from New South Wales, established a short-lived settlement at Risdon Cove, on the Derwent River, in 1803. *Collins, sent from England, abandoned an attempt to settle on the coast of Port Phillip, near present-day *Sorrento, and established a settlement at Sullivan's Cove, also on the Derwent River, in 1804. He named it Hobart Town. Collins was the first Lieutenant-Governor of Van Diemen's Land. William *Paterson, sent from New South Wales, established a settlement at Port Dalrymple, in the north of the island, in late 1804. Two years later he moved the settlement to the site of present-day Launceston. The settlements in the north and south were administered separately until 1812. Van Diemen's Land was separated administratively from New South Wales in 1825. Convicts were sent to Van Diemen's Land, originally from New South Wales and then directly from Britain, until 1853. *Responsible government was proclaimed in 1855, the first responsible ministry taking office the following year. The colony's new name, Tasmania, was proclaimed in 1855.

VAN DIEMEN'S LAND COMPANY An agricultural and pastoral organization, formed by a London syndicate in 1825. Its first agent, Edward Curr, chose land in *Van Diemen's Land the following year. The company was eventually granted about 141 000 hectares. Its major activities continue to be agricultural and pastoral ones.

VERSAILLES, TREATY OF A treaty between the Allies and Germany in 1919. It was one of a number of treaties whose provisions were determined at the *Paris Peace Conference, which was held at the end of the First World War. It redrew the boundaries of European countries; redistributed German colonial possessions; determined the war reparations to be paid by Germany; and established the *League of Nations, among other provisions. The Australian Commonwealth Parliament ratified the Treaty in 1919. Australia was granted a mandate by the League of Nations to administer former German *New Guinea and associated islands south of the equator; a share, with Britain and New Zealand, in a mandate over Nauru; and reparations from Germany, of which £5 571 720 (much less than Australia was promised, which in turn was much less than she had claimed) ultimately was paid before the system collapsed in 1932. Australia also became a member of the League of Nations.

VICTORIA, a state of the *Commonwealth of Australia, is located in the south-east of the continent. In its early years of White settlement it was known as the *Port Phillip District and formed part of New South Wales. Early convict settlements there were short-lived (1803–4, 1826–8). Although *sealers and *whalers visited, the first permanent White settlements did not begin until 1834 and 1835. The area was separated

Anti-war pamphlet, Mothers in Mourning *(1970), issued by the Save Our Sons Movement during the* **Vietnam War.** *(La Trobe Collection, State Library of Victoria)*

MOTHERS IN MOURNING

Commemorate Mother's Day, 1970, in a different way. Cry halt to the Vietnam tragedy! Join the Vietnam Moratorium action to end the War and bring our boys home.

For details write to Vietnam Moratorium Campaign, First Floor, 107 Liverpool Street, Sydney. Phone 61-2522. Watch the press for details.

Authorised for Save Our Sons Movement by Mrs Adele Pert, 25 Anzac Ave, Ryde. Printed by Comment Publishing Company, 22 Steam Mill Street, Sydney.

from New South Wales, and named Victoria, in 1851. *Responsible government was proclaimed in Victoria in 1855, and Victoria became a state of the newly created Commonwealth of Australia in 1901. *Melbourne is the capital city of Victoria.

VICTORIA (north Australia) *Port Essington

VIETNAM WAR This term generally refers to the period of conflict between North Vietnam, supported by China and the Soviet Union, and South Vietnam, supported by the USA and other countries, culminating in victory for the North Vietnamese in 1975. Australia, as a member of the *South-East Asia Treaty Organization, sent a small number of army instructors to support South Vietnam in 1962 and 1964, and then began to send combat forces in 1965. Harold *Holt, the Australian Prime Minister, used a now-famous phrase, 'All the way with LBJ', to pledge Australia's support for the USA, in general and in relation to its participation in the Vietnam War, in a speech to the President, L. B. Johnson, during a visit to the United States in mid-1966. The full sentence was:

And so, sir, in the lonelier and perhaps even more disheartening moments which come to any national leader, I hope there will be a corner of your mind and heart which takes cheer from the fact that you have an admiring friend [Australia], a staunch friend that will be all the way with LBJ.

By 1968 there were more than 8000 Australian men, from the three armed services, in Vietnam. They included conscripts. The main area of Australian involvement was Phuoc Tuy Province. The Australians were withdrawn from Vietnam during 1972, by which time some 47 000 Australians had served there. Just under 500 Australians had been killed, and approximately 2400 had been wounded. During the late 1960s and early 1970s, Australia's involvement in the Vietnam War, and the *conscription issue, became the focus of bitter public debates.

VLAMINGH PLATE *Hartog, Dirck

VOYAGER DISASTER Australia's worst naval peacetime disaster. The destroyer HMAS *Voyager* and aircraft-carrier HMAS *Melbourne* collided just before 9 p.m. on 10 February 1964 during night exercises off Jervis Bay, New South Wales, after the *Voyager* turned off course into the *Melbourne*'s path. The *Voyager* was cut in half. Its bow sank immediately, its stern by about midnight. Killed were 82 members of its crew of 324, including its commander, Captain Duncan Stevens, and all of its other bridge officers. The *Melbourne*'s bow was damaged, but its crew suffered no casualties. A royal commission held in 1964 found that the *Voyager* caused the collision by turning off course, but also criticised the *Melbourne*'s commander, Captain John Robertson, who was in overall charge of the exercises, for failing to take avoiding action. The Naval Board did not blame Robertson, but transferred him to a shore posting. He resigned from the navy in September 1964. Continuing controversy, including allegations about Stevens's drinking, led to a second royal commission in 1967 and 1968. It found that the *Voyager* was solely responsible for the collision, that Stevens was unfit to command the *Voyager* because of ill health, and that criticism of Robertson was unjustified. Robertson received some financial compensation from the government in 1968.

W

WAKEFIELD, EDWARD GIBBON (1796–1862), colonial theorist, was born in London, and was educated at Westminster School and Edinburgh High School, before pursuing a diplomatic career. While serving a sentence in Newgate Gaol (1827–30) resulting from his abduction of a fifteen-year-old heiress, he began to develop theories on colonization, which eventually covered such aspects as penal reform, land policies, migration, and forms of government. He is remembered as the chief exponent of *systematic colonization. Wakefield has been credited, some would say inaccurately, with having influenced migration and land policies for *New South Wales; the ending of *transportation to New South Wales; the colonization of *South Australia, *Western Australia (notably at *Australind), and *New Zealand; and colonial government policy, through Lord Durham's *Report on the Affairs of British North America* (1839). He did not visit any of the Australian colonies, but did live for periods in Canada and New Zealand. Wakefield's numerous publications (some of which were published anonymously or under other names) include *A Letter from Sydney* (1829), *Sketch of a Proposal for Colonizing Australasia* (1829), *England and America* (1833), and *A View of the Art of Colonization* (1849).

WALLACE'S LINE The name given to an imaginary line, running between Bali and Lombok, said to divide the fauna of Asia from that of Australasia. T. H. Huxley named it in 1868, after Alfred Russel Wallace, an English scientist, who had earlier made this distinction.

'WALTZING MATILDA' One of Australia's most popular songs. In 1895, while staying on a station near Winton, in Queensland, 'Banjo' *Paterson wrote the words, which tell the story of a *swagman's adventures ('waltzing matilda' means carrying a swag). The tune is an adaptation of a Scottish song named 'Thou Bonnie Wood of Craigielea'. 'Waltzing Matilda', which was first sung publicly in a Winton hotel in April 1895, has periodically (but unsuccessfully) been proposed as Australia's national anthem.

WARBURTON, PETER EGERTON (1813–89), was born in England, and joined the Royal Navy at the age of twelve. He attended the Royal Indian Military College, Addiscombe, Surrey, from 1829 to 1831, and then served in the Bombay Army until 1853, when he retired with the rank of major. During his term of office as the South Australian Commissioner of Police, which lasted from 1853 until 1867, Warburton explored a number of parts of the colony's interior. He became chief staff officer and colonel of South Australia's Voluntary Military Force in 1869, remaining in the force until 1877. Warburton left Adelaide in September 1872 in charge of an expedition, financed by Thomas Elder and Walter Hughes, to explore from central Australia to the west. The party, comprising Warburton, his son Richard, two other Europeans, two Afghan cameldrivers and an Aboriginal tracker, left Alice Springs in April 1873, and had to cope with harsh country, lack of water, extreme heat, and illness, and eventually had to eat some of their seventeen camels. Warburton was obliged to abandon his early plan of aiming for Perth. After almost perishing, the party finally arrived at Roebourne in north-western Western Australia in January 1874, becoming the first to cross Australia from the centre to the west. Warburton, who had lost the sight of one eye on the expedition, was awarded the Royal Geographical Society's gold medal, and his *Journey Across the Western Interior of Australia* was published in 1875. He died at his property near Adelaide.

WARTIME TRANSPORT STRIKE A major industrial dispute in 1917, and the largest strike in Australia to that time. Workers in the New South Wales Railways and Tramways Department, protesting over the introduction of a 'card system' to record the time taken for each job, which they saw as a form of 'speeding up', went on strike on 2 August. Waterside workers, coal-miners, and others soon withdrew their labour in support of them. About 97 500 Australian workers (of whom about 76 000 were in New South Wales) went on strike. The New South Wales Nationalist Government and other employers used so-called 'loyalist' labour to maintain essential industries. The railway workers returned to work in September. The 'card system' remained, a royal commission was promised within three months, and many workers were demoted. All remaining strikers had returned to work before the end of October.

WATSON, JOHN CHRISTIAN (1867–1941), politician, was born in Valparaiso, Chile. His parents, who were English, were travelling to New Zealand, where Watson grew up and worked in the printing trade. He moved to Sydney when he was nineteen, worked as a compositor, and was active in the Typographical Union. Watson became President of the Sydney Trades and Labour Council, and of the Australian Labour Federation. He was a Labor Member of the Legislative Assembly in New South Wales from 1894 to 1901. He was then a Member of the House of Representatives from 1901 to 1910, and led the federal parliamentary Labor Party from 1901 to 1907. He became the first, albeit short-lived, Labor Prime Minister and Treasurer (1904). In 1916 Watson was expelled from the Labor Party for supporting *conscription.

WAVE HILL A Northern Territory pastoral station that was the scene of significant events in the Aboriginal *land rights movement. In 1966 Gurindji people at Wave Hill station, which was leased by a large British firm, went on strike over low wages and poor conditions, and occupied station land at Wattie Creek. In April 1967 they petitioned the Governor-General, asking for some 1300 square kilometres of their traditional land around Wattie Creek. The petition was rejected. In 1975 the Gurindji people were granted leasehold title to about 3280 square kilometres of the land, known as Daguragu; an Aboriginal land trust has since been granted freehold title to most of Daguragu.

WCTU *Woman's Christian Temperance Union

'WELCOME STRANGER' The name given to the largest gold nugget found in Australia. John Deason and Richard Oates found the nugget at Moliagul, near Dunolly, Victoria, in 1869. Its gross weight was 78 381 grams; its net weight was 71 040 grams. Deason and Oates received £9534 for it.

WENTWORTH, WILLIAM CHARLES (1790–1872), politician, was the son of Catherine Crowley, a convict, and D'Arcy Wentworth, a surgeon. His parents had both sailed to New South Wales with the *Second Fleet. Wentworth lived on

*Norfolk Island and in New South Wales before being educated in England between 1803 and 1810. He returned to New South Wales and became the Acting Provost-Marshal in 1811. Blaxland, Lawson, and Wentworth found a route that led to the crossing of the *Blue Mountains in 1813. Wentworth worked on a schooner for some time before returning to England in 1816. He won second prize with his poem 'Australasia' in a competition for the Chancellor's gold medal while he was at Peterhouse, Cambridge. Wentworth pursued a legal career; he entered the Middle Temple in 1817, and was called to the Bar in 1822. His *A Statistical, Historical, and Political Description of the Colony of New South Wales and its dependent Settlements in Van Diemen's Land, With a Particular Enumeration of the Advantages which these Colonies offer for Emigration and their Superiority in many Respects over those possessed by the United States of America* (1819) was influential. In 1824 Wentworth returned to New South Wales, where he led the *emancipists, campaigned for political and legal reform, established the *Australian* newspaper with Robert Wardell, and continued his legal career. He became involved in the *Sudds and Thompson affair, and was active in the *Australian Patriotic Association. He also became a *squatter with large holdings. In 1840 Wentworth and some associates tried to buy about one-third of New Zealand, but were foiled in the attempt by *Gipps. Wentworth played a part in drawing up the draft bill forming the basis for the Act of 1842 that established a partly-elected, partly-nominated Legislative Council in New South Wales. He entered the Legislative Council in 1843, and led the opposition to Gipps, especially over the latter's attempts to control the squatters. Wentworth, who had become increasingly conservative, chaired the Select Committee that drafted the New South Wales Constitution in 1853. He and Edward Deas Thompson, the Colonial Secretary, went to England, where they saw it become law in 1855. Wentworth spent the rest of his life in England, except for one brief period in 1861–2 when he returned to New South Wales and was President of the Legislative Council.

WESTERN AUSTRALIA is a state of the *Commonwealth of Australia. *Hartog, a Dutch mariner, is the first European known to have landed on the west coast of Australia, in 1616. *Dampier, an English explorer, visited the north-west coast in 1688 and 1699. Early reports were unfavourable. In 1826 Governor *Darling, acting on British instructions, sent soldiers and convicts to King George Sound, where they established a settlement named Frederickstown, later renamed *Albany. In 1829 Britain annexed Australia's western third and founded the free colony of Western Australia. The new colony's main settlement was established at *Perth, on the Swan River. Some ten thousand convicts, all male, were transported to Western Australia between 1850 and 1868. *Responsible government was proclaimed in Western Australia in 1890, and the colony became a state of the newly created Commonwealth of Australia in 1901.

WEST IRIAN *Irian Jaya

WEST GATE BRIDGE A bridge across Melbourne's Yarra River, some 3 kilometres from its mouth, linking the south-western suburbs with the city. During construction, which had begun under

William Duke's The Rounding, *showing* **whalers** *at work, was published in Hobart Town in 1848. Van Diemen's Land was an important whaling base. (W. L. Crowther Library, State Library of Tasmania)*

the Lower Yarra Crossing Authority's control in April 1968, a steel span on the western side of the bridge collapsed on 15 October 1970. Thirty-five workers, some of whom were on the span and others in huts below, were killed, and several others were seriously injured. A royal commission, which began in 1970, found on 3 August 1971 that the designers, the London firm of Freeman, Fox and Partners, were largely responsible for the disaster, but also blamed the Authority, contractors, trade unions, and workers. Work on a modified design resumed in February 1972 and the bridge was finally opened on 15 November 1978.

Westminster, Statute of A British Act of Parliament in 1931 that gave equal status to the self-governing dominions (including Australia) of the British Empire. It was based on resolutions of the Imperial Conferences of 1926 and 1930. In 1926 Lord Balfour's committee had defined the UK and dominions as:

> *autonomous communities within the British Empire, equal in status, in no way subordinate to one another in any aspect of their domestic or external affairs, though united by a common allegiance to the Crown and freely associated as members of the British Commonwealth of Nations.*

Australia ratified the *Statute of Westminster* in 1942.

Westpac Banking Corporation *Bank of New South Wales

Whalers from Australia, Britain, America, and other places operated around Australia from the 1790s. Whales were hunted for oil and whalebone, and whale and seal products were Australia's main exports until the 1830s. Numerous whaling stations were established in Van Diemen's Land (which had forty-one bay-whaling stations by 1841), New South Wales, the Port Phillip District, South Australia, and Western Australia. Whaling declined during the latter half of the nineteenth century, but continued throughout much of the twentieth century. Australia's last whaling station, in Western Australia, closed in 1978.

WHITE, PATRICK VICTOR MARTINDALE (1912–90), writer, was born to Australian parents in London. His father was a grazier. White was educated at Tudor House in Moss Vale, New South Wales, Cheltenham College in England,

*Photograph of Patrick **White**, one of Australia's best-known writers, taken by Eva Pavlovic in 1970. (Eva Pavlovic)*

and King's College, Cambridge. Before going up to Cambridge, White jackarooed at Monaro and Walgett; after coming down, he travelled in Europe and the USA. During the Second World War he was an intelligence officer with the Royal Air Force in the Middle East and Greece. He returned permanently to Australia in 1948. White won the Nobel Prize for Literature in 1973. His novels include *Happy Valley* (1939), *The Living and the Dead* (1942), *The Aunt's Story* (1946), *The Tree of Man* (1955), *Voss* (1957), *Riders in the Chariot* (1961), *The Solid Mandala* (1966), *The Vivisector* (1970), *The Eye of the Storm* (1973), *A Fringe of Leaves* (1976), *The Twyborn Affair* (1979), and *Memoirs of Many in One* (1986). His other publications include stories, poetry, plays, a screenplay, and an autobiography.

WHITE AUSTRALIA POLICY A restrictive *immigration policy pursued by Australia for many years. All Australian colonies restricted Asian (particularly Chinese) immigration at times during the second half of the nineteenth century. In 1901 the new Commonwealth Parliament legislated to exclude non-Europeans. The main device used was a dictation test in any European language (any 'prescribed' language from 1905). This policy of exclusion continued, virtually unchanged, until the 1950s. Entry permits, based largely upon economic criteria, replaced the dictation test in 1958.

WHITLAM, EDWARD GOUGH (1916–), politician, was born in Kew, Victoria, and educated at Knox Grammar School in Sydney, Telopea Park High School in Canberra, Canberra Grammar School, and the University of Sydney, from which he graduated in Arts in 1938, and Law in 1946. His father was a lawyer. During the Second World War Whitlam served with the Royal Australian Air Force from 1941 to 1945. He was admitted to the New South Wales and Federal Bars in 1947, and was appointed Queen's Counsel in 1962. He was a New South Wales Labor Member of the House of Representatives from 1952 to 1978, led the Opposition twice (1967–72, 1975–7), and was Prime Minister from 1972 to 1975, and Minister for Foreign Affairs from 1972 to 1973. The Whitlam Government's achievements, some of which were controversial, include the abandoning of *conscription and withdrawal of Australian troops from the *Vietnam War, the granting of diplomatic recognition to the Chinese People's Republic, the abolition of university fees, and the introduction of a national health system known as Medibank. The *dismissal of the Whitlam Government in 1975 caused great controversy. After Labor's defeat in the 1977 elections, Gough Whitlam resigned from his party's parliamentary leadership, and from parliament. He was a Visiting Fellow at the Australian National University from 1978 to 1979, that institution's first National Fellow from 1980 to 1981, and Australia's Ambassador to Unesco from 1983 to 1986. His numerous publications include *The Truth of the Matter* (1979) and *The Whitlam Government* (1985).

WILLIAMSON, JAMES CASSIUS (1845–1913), actor and entrepreneur, was born in Pennsylvania, and gained theatrical experience in various places, including New York and San Francisco, before he and his first wife, Maggie Moore, having been engaged by George *Coppin, appeared in *Struck Oil* in Australia in 1874 and 1875. After touring elsewhere

Gough **Whitlam**, seen here in a photograph taken by J. Crowther, was Australia's Prime Minister from 1972 to 1975. (Australian Information Service)

with the play, they returned to Australia in 1879 with rights to *H.M.S. Pinafore*. Williamson turned increasingly to management, forming a comic-opera company in 1880, and another organization (originally in partnership with George Musgrove and Arthur Garner), which eventually became known as 'the Firm', in 1882. 'The Firm' presented many plays, comic operas and musical comedies, mostly English and American, in Australia and New Zealand. Williamson, who has been described as 'Australia's most successful theatrical entrepreneur', died in Paris. The organization (which had become J. C. Williamson Limited in 1907) continued for many years after his death.

'WOBBLIES'

Lithographic poster, printed by Troedel & Co., Melbourne, for Struck Oil, *starring J. C.* **Williamson** *and Maggie Moore. (La Trobe Collection, State Library of Victoria)*

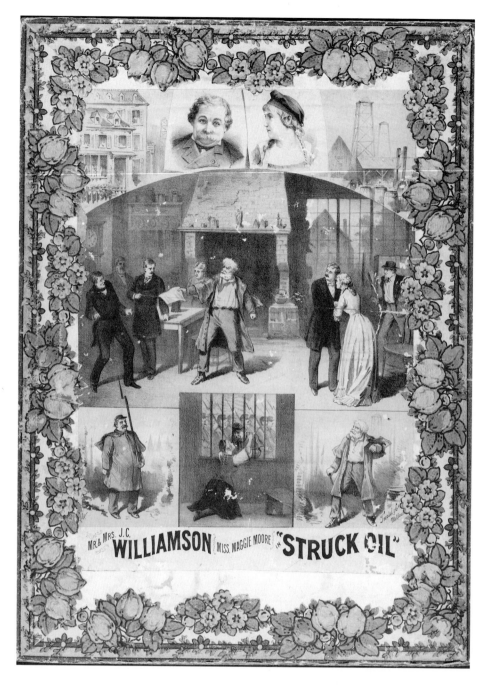

'WOBBLIES' *Industrial Workers of the World

WOLSELEY, FREDERICK YORK (1837–99), inventor, was born in Ireland, arrived in Melbourne in 1854, and worked on, and eventually owned, various sheep stations in eastern Australia. He experimented with sheep-shearing machines, taking out several patents (one with R. Savage, another with R. P. Park) for such devices in 1877 and 1884. John Howard (who sold the rights to his horse-clipper to Wolseley, and was employed by him)

improved them. A number of Australian shearing sheds began to use Wolseley's shearing-machine in 1888. In 1889 Wolseley went to England, founded the Wolseley Sheep Shearing Machine Co. Pty Ltd in Birmingham, and was its managing director until 1894. (The firm later also manufactured Wolseley motor cars.) He died in London.

WOMAN'S CHRISTIAN TEMPERANCE UNION (WCTU) An organization of women with branches in each state. The organization was founded in the United States of America in 1874, then spread to other countries. The first Australian branch began in Sydney in 1882, and branches in other colonies followed during the 1880s. The WCTU's primary objective has been the restriction, and eventual abolition, of the use of alcoholic beverages. The Woman's Christian Temperance Union played an active role in the campaign for female suffrage in Australia, although some historians have suggested that this role may have been over-emphasized.

WOOMERA is a town in South Australia, about 180 kilometres north-west of Port Augusta. It lies within the Woomera rocket range, which extends over a large area in South Australia and Western Australia, and to which public access is restricted. Both were developed as a result of the United Kingdom – Australia Long Range Weapons Agreement (also known as the Joint Project Agreement) of 1947. (A woomera is an Aboriginal spear-throwing stick.) Various countries have been involved in programmes conducted at the range. Maralinga and Emu,

*Photograph of delegates at the 1898 annual conference of the **Woman's Christian Temperance Union** held at the Congregational Church in Melbourne, published in the* Weekly Times, *3 December 1898. (La Trobe Collection, State Library of Victoria)*

295

two of the sites of British *nuclear tests in Australia, lie within the area, as does the United States Joint Defence Space Communications Station at Nurrungar.

'WORKINGMAN'S PARADISE' Popular term used to describe Australia, particularly during the second half of the nineteenth century. Henry Kingsley first used the term sarcastically to describe Australia in his novel *Recollections of Geoffry Hamlyn* (1859), but it was frequently misquoted, many people taking it literally. William *Lane used the term, with irony, as the title of his socialist novel *The Working Man's Paradise: An Australian Labour Novel* (1892).

WORLD WAR I *First World War

WORLD WAR II *Second World War

WYBALENNA *Furneaux Islands

Z

ZEEHAN

Rudolph W. W. Koch's lithograph Tasmanian Smelting Company Limited, General View of Works *(1899), showing the smelting works at Zeehan in 1899. (La Trobe Collection, State Library of Victoria)*

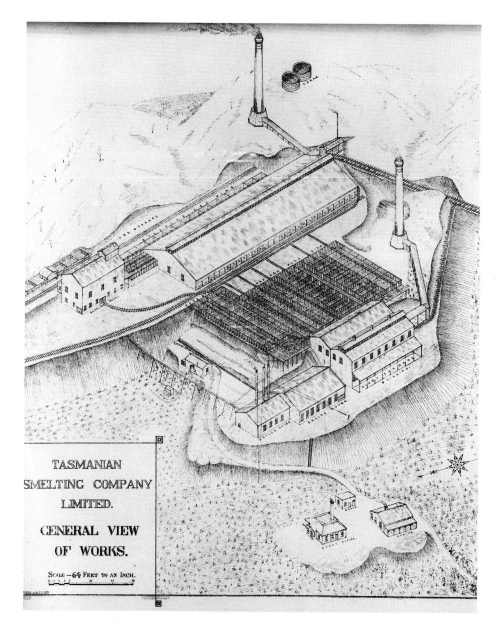

ZEEHAN, a town in western Tasmania, lies some 5 kilometres to the north of Mount Zeehan, a peak sighted in 1642 by *Tasman, the Dutch mariner. (*Flinders later named the peak after the *Zeehaen*, one of Tasman's ships.) Silver-lead deposits were discovered in the area in 1882, leading to the establishment of Zeehan. It developed rapidly during the 1890s, becoming a town in 1898 and a municipality in 1906, and its population reputedly reached about 8000 in the early 1900s. Mining in the area (and consequently the town itself) declined before the First World War, but a new lode of ore was discovered in 1949. During the 1960s the renewal of tin mining at Renison Bell, 17 kilometres to the north-east, also assisted Zeehan.

ILLUSTRATIONS AND ACKNOWLEDGEMENTS

Page 3
Charles Bayliss 1850–1897 Australian
Group of Aborigines at Chowilla Station, Lower Murray, SA
Albumen silver photograph
23.3 x 29.3 cm
Purchased through the Art Foundation of Victoria, 1992
National Gallery of Victoria

Page 6
Sidney Nolan
Kenneth Soldier 1958
Polyvinyl acetate on hardboard
122 x 91.5 cm
19577
Australian War Memorial

Page 9
Australia Day pamphlet, 1918
Australian Natives' Association

Page 12
Photograph of men of the original Light Horse of the AIF before departure
J450
Australian War Memorial

Page 14
Australian Labor Party poster, 1928
Copy in Author's Collection

Page 15
top
Menu for ANA Luncheon, 1935
Australian Natives' Association
bottom
Kaz Cooke
PSST! What's the Capital of Argentina?
H90.1614
La Trobe Collection
State Library of Victoria

Page 16
Photograph of AWM
X125
Australian War Memorial

Page 18
Photograph of Bank of New South Wales
Westpac Banking Corporation

Page 19
Trial of Edward Kelly
from *Australasian Sketcher*, 20 November 1880
National Library of Australia

Page 20
Photograph of Edmund Barton
Australian Natives' Association

Page 22
Photograph of the *Batavia* taken by Patrick Baker
MA 3148A
Western Australian Maritime Museum

Page 23
A. G. Bolam
Daisy Bates at Ooldea 1920
H33975A SPF
La Trobe Collection
State Library of Victoria

Page 25
Photograph on *Jervis Bay*, 1925
Big Brother Movement

Page 27
Blackbird Taming
from *Illustrated Melbourne Post*, 1872
National Library of Australia

Page 28
William Strutt
Black Thursday 1862–4
H28049
La Trobe Collection
State Library of Victoria

Page 29
Mick Armstrong
Now bring out your Larwood
from Melbourne *Herald*, 25 November 1932
La Trobe Collection
State Library of Victoria

Page 30
Patriotic Nurses for Boer War
AP476/14
Australian Archives, South Australia

Page 32
A map of Port Jackson by J. Walker
ML ref M2811.1/1791/2
Mitchell Library
State Library of New South Wales

Page 33
1829 lithograph of Sir Richard Bourke
National Library of Australia

Page 34
Photograph of Edward Braddon
Australian Natives' Association

Page 35
Photograph of Don Bradman
The Herald & Weekly Times Ltd

Page 36
Queen Street, Brisbane, in 1860
H8482 SPF
La Trobe Collection
State Library of Victoria

Page 38
Broken Hill strike leaflet
National Library of Australia

Page 40
S. E. Campbell
Frontispiece for Mary Grant Bruce, *A Little Bush Maid*, undated reprint
Ward Lock Ltd

Page 41
Photograph of S. M. Bruce
Copy in Author's Collection

Page 42
W. Macleod
William Buckley
from A. Garran (ed.), *Picturesque Atlas of Australasia*, vol. 1, 1886
H21541 SPF
La Trobe Collection
State Library of Victoria

Page 43
The Workman's Cross: Disunion
from *Bulletin*, 30 May 1891
La Trobe Collection
State Library of Victoria

Page 44
The Departure of the Burke and Wills Expedition 1881
Coloured lithograph from the supplement to the *Illustrated Australian News*, May 1881
National Library of Australia

Page 45
Bushranging Vagaries in New South

Illustrations and Acknowledgements

Wales from *Melbourne Punch*, 22 October 1863
National Library of Australia

Page 46
John Rowell
She set out for the shore
The Victorian Readers

Page 50
Unknown
Convict Uprising at Castle Hill 1804
Rex Nan Kivell Collection
National Library of Australia

Page 52
Photograph of Joseph Benedict Chifley
National Library of Australia

Page 53
J. W. Beattie
Andrew Inglis Clark the elder
from J. W. Beattie, *Members of the Parliaments of Tasmania* c. 1900
Allport Library and Museum of Fine Arts
State Library of Tasmania

Page 55
Advertisement for film *For the Term of His Natural Life*
Archives Office of Tasmania

Page 58
Photograph of Victoria Mill, 1883
Archives negative 1581
Deposit 171/733/4
Archives of Business and Labour
Australian National University

Page 62
'The Blood Vote'
Anti-conscription leaflet
A4801
Australian War Memorial

Page 65
S. Calvert
Coppin's Olympic Theatre 1872
H4991 MFN 239
La Trobe Collection
State Library of Victoria

Page 67
E. Gilks
This Sketch of the Victorian Eleven and the Intercolonial Cricket Ground, Melbourne, Feby. 2–3 & 4— 1860
H2084 LT820 SPF
La Trobe Collection
State Library of Victoria

Page 69
Photograph of John Curtin
National Library of Australia

Page 71
Richard Daintree 1832–1878 Australian
Darley Quarries, Bacchus Marsh, Victoria c. 1858
Albumen silver photograph
16.8 x 22.5 cm

Purchased 1992
National Gallery of Victoria

Page 74
'Dollar Bill'
Treasury, Canberra

Page 76
Illustration by Hal Gye from C. J. Dennis, *The Glugs of Gosh*, 1917
Collins Angus & Robertson Publishers

Page 78
Ways and Means
National Library of Australia
Reproduced with the permission of the Royal Commonwealth Society

Page 79
Distributing the Food at the So. Melbourne depot
from *Leader*, 14 April 1894
La Trobe Collection
State Library of Victoria

Page 80
William Dobell 1899–1970
Australian
Two Men Carrying a Load c. 1943
Pen and ink
12.6 x 12.5 cm
Purchased 1946
National Gallery of Victoria
Reproduced with the permission of Tony Clune

Page 82
Russell Drysdale 1912–1981
Australian
Greek Refugees before Bombed Buildings in Larissa, Greece 1943
Pen and ink
20.1 x 23.5 cm
Purchased 1944
National Gallery of Victoria
Reproduced with the permission of Lady Drysdale

Page 83
Will Dyson
The Wine of Victory—German Prisoners—The Salient 1919
Lithograph
46.4 x 59.5 cm
2293
Australian War Memorial

Page 87
The Eight Hours Annual Celebration
from *Australasian Sketcher*, 14 June 1873
La Trobe Collection
State Library of Victoria

Page 90
Edward John Eyre 1868
SSL:M:B8429
Mortlock Library of Australiana
State Library of South Australia

Page 93
Mick Armstrong
'*Hulloa, Caretaker!*'

H35334
La Trobe Collection
State Library of Victoria

Page 96
J. S. Watkins
Women of Queensland!
Lithograph
99 x 72 cm
V5632
Australian War Memorial

Page 97
William Westall 1781–1850
Australian
A View of King George's Sound 1802
Watercolour
27.9 x 42.9 cm
Purchased 1978
National Gallery of Victoria

Page 100
1928 VFL Grand Final scenes
from *Australasian*, 6 October 1928
La Trobe Collection
State Library of Victoria

Page 101
Photograph of John Forrest
Australian Natives' Association

Page 103
Photograph of Malcolm Fraser
Age

Page 108
Unknown
Laying the Foundation Stone of Geelong Clock Tower 1856
Ambrotype
6.5 x 5.5 cm
Purchased 1974
National Gallery of Victoria

Page 110
William Dobell 1899–1970
Professor L. F. Giblin 1945
Pencil drawing
26.6 x 17.7 cm
National Library of Australia
Reproduced with the permission of Tony Clune

Page 111
S. T. Gill 1818–1880 Australian
Native Sepulchre
Watercolour
29.9 x 46.1 cm
Puchased 1972
National Gallery of Victoria

Page 112
William Dobell 1899–1970
Dame Mary Gilmore 1957
Oil on hardboard
90.2 x 73.7 cm
Gift of Dame Mary Gilmore 1960
Art Gallery of New South Wales

Page 114
John Glover 1767–1849
Australian

Illustrations and Acknowledgements

The River Nile, Van Dieman's [sic] Land 1837
Oil on canvas
76.2 x 114.3 cm
Felton Bequest 1956
National Gallery of Victoria

Page 115
S. T. Gill 1818–1880 Australian
The New Rush 1863
Hand coloured lithograph
28.8 x 42.6 cm
Purchased 1954
National Gallery of Victoria

Page 116
Vida Goldstein portrait, c. 1902
MS 6666 Box 332/3
La Trobe Collection
State Library of Victoria

Page 118
Photograph of John Gorton with Miss Australia, March 1969
Federal Capital Press of Australia Pty Ltd

Page 120
Photograph of Percy Grainger, 1934
Grainger Museum
The University of Melbourne

Page 121
Ted Scorfield
'You're Next!'
from *Bulletin*, 2 July 1930
National Library of Australia

Page 122
Great White Fleet postcard
Edith Coels Papers
Author's Collection

Page 124
Phil May
Dishing his Enemies 1888
H30382
La Trobe Collection
State Library of Victoria

Page 129
Vlamingh Plate
National Library of Australia

Page 131
Tom Roberts 1856–1931 Australian
The Artists' Camp c. 1886
Oil on canvas
115.8 x 73.7 cm
Felton Bequest
National Gallery of Victoria

Page 133
Photograph of Henry Bournes Higgins
Australian Natives' Association

Page 135
A. C. Cooke
Hobart Town 1879
Black and white wood engraving
from supplement to the *Australasian Sketcher*, 1879
National Library of Australia

Page 136–7
Bernhardt Holtermann Australian
Panorama of Sydney Harbour and Suburbs 1875
Albumen silver photograph
38.9 x 310.2 cm
Purchased through the Art Foundation of Victoria 1990
National Gallery of Victoria

Page 138
'Billiwog' cartoon
from *The Billy Book, Hughes Abroad, Cartoons by Low* 1918

Page 139
An oil painting by William Minehead Bennett of Vice Admiral John Hunter
ML ref ZDG 394
Dixson Galleries
State Library of New South Wales

Page 141
Frank Hurley 1885–1962 Australian
We took a dog team down to the wreck and salvaged a few essentials 1915
Gelatin silver photograph
56.6 x 41.3 cm
Australian National Gallery, Canberra

Page 142
'The Return of Prosperity'
from *Illustrated Tasmanian Mail*, 11 November 1931
Archives of Tasmania

Page 144
Marshall Claxton 1812–1881 Australian
An Emigrant's Thoughts of Home 1859
Oil on board
60.5 x 46.8 cm
Presented by the National Gallery Women's Association 1974
National Gallery of Victoria

Page 146
Photograph of Isaac Isaacs
Australian Natives' Association

Page 148
Jackaroo illustration
Copy in Author's Collection

Page 152
Spooner
High Interest Rates 1989
H89.250
La Trobe Collection
State Library of Victoria

Page 153
Advertising poster for *The Story of the Kelly Gang*
National Library of Australia

Page 154
Unknown
An oil painting of Governor King
ML ref ZML 546
Mitchell Library
State Library of New South Wales

Page 155
Captain Kingsford Smith and C. J. P. Ulm
Postcard
H8795
La Trobe Collection
State Library of Victoria

Page 156
Photograph of Charles Kingston
Australian Natives' Association

Page 157
Tom V. Carter
An Errand of Mercy
The Victorian Readers

Page 159
Eureka!
H81.86/7 SPF
La Trobe Collection
State Library of Victoria

Page 161
Photograph of William Lane
Archives negative 1563
Deposit E97/26
Archives of Business and Labour
Australian National University

Page 162
Photograph of J. T. Lang
The Herald & Weekly Times Ltd

Page 163
The Youthful Larrikin
from Harry Furniss, *Australian Sketches*
National Library of Australia

Page 164
Photograph of John Latham
Copy in Author's Collection

Page 165
Francis Grant
Charles Joseph La Trobe 1855–6
H38452 SPF
La Trobe Collection
State Library of Victoria

Page 166
Frederick Strange
Brisbane Street, Launceston 1858
Queen Victoria Museum and Art Gallery, Launceston

Page 167
Henry Lawson
'Just Like Home'
National Library of Australia

Page 170
Photograph of Norman Lindsay
ML ref P1/L
Mitchell Library
State Library of New South Wales

Page 172
Photograph of Joe Lyons
National Library of Australia

Page 174
Joseph Lycett
The Residence of John McCarthur Esqre. near Parramatta, New South Wales

301

Illustrations and Acknowledgements

Page 176
    National Library of Australia
Page 176
    Frederick McCubbin   1855–1917
        Australian
    Lost   1886
    Oil on canvas
    115.8 x 73.7 cm
    Felton Bequest 1940
    National Gallery of Victoria
Page 177
    Photograph of Dorothea Mackellar,
        c. 1915–16
    ML ref P1/M
    Mitchell Library
    State Library of New South Wales
Page 178
    Portrait of Mother Mary McKillop
    Sisterhood of St Joseph of the Sacred
        Heart
Page 179
    A miniature of Major General Lachlan
        Macquarie
    ML ref ZMIN 71
    Mitchell Library
    State Library of New South Wales
Page 180
    Michael Leunig
    The Wreck
    from *Age*, 4 April 1992
    Age
Page 183
    Photograph of Nellie Melba by Arnold
        Genthe
    Ltaf 295.f.15
    La Trobe Collection
    State Library of Victoria
Page 184
    Collins Street at Four P.M.
    from *Illustrated Australian News*, 18
        July 1868
    La Trobe Collection
    State Library of Victoria
Page 185
    The Cup Day: A Sketch on the Lawn
    from *Australasian Sketcher*, 29
        November 1873
    La Trobe Collection
    State Library of Victoria
Page 186
    Photograph of Robert Gordon Menzies
    National Library of Australia
Page 189
    Walter G. Mason
    T. S. Mort Esq., of Sydney
    Wood engraving
    Rex Nan Kivell Collection
    National Library of Australia
Page 190
    Portrait of Baron von Mueller
    H5057 LT 775 SPF
    La Trobe Collection
    State Library of Victoria

Page 191
    Photograph of Sidney Myer's Bendigo
        drapery store
    Myer Stores Ltd
Page 195
    David Low
    Patching the Drum
    H32945
    La Trobe Collection
    State Library of Victoria
Page 196
    Thomas Ham
    Native Police Encampment
    La Trobe Collection
    State Library of Victoria
Page 197
    New Australian Colony—Cosme, Paraguay
    National Library of Australia
Page 200
    Photograph of church at New Norcia
    73713P
    J. S. Battye Library of West Australian
        History
Page 201
    A Flood in New South Wales
    National Library of Australia
Page 203
    A chain gang on Norfolk Island
    National Library of Australia
Page 208
    S. T. Gill
    Overlanders c. 1865
    Watercolour on paper
    33.5 x 56 cm
    Purchased 1946
    Art Gallery of New South Wales
Page 209
    An Interruption on the Overland
        Telegraph Line
    H34959 SPF
    La Trobe Collection
    State Library of Victoria
Page 211
    Photograph of Earle Page
    National Library of Australia
Page 212
    Phil May
    The Mother of Civilization 1888
    H30387
    La Trobe Collection
    State Library of Victoria
Page 216
    Photograph of Mrs Petrov, 1954
    News Limited
Page 217
    Photograph of Melbourne Cup, 1930
    H19100 SPF
    La Trobe Collection
    State Library of Victoria
Page 218
    Engraving by W. Sherwin, from a
        painting by F. Wheatley of

        Arthur Phillip
    Rex Nan Kivell Collection
    National Library of Australia
Page 219
    Photograph taken during Melbourne
        Police Strike
    National Library of Australia
Page 220
    Photographs of convicts at Port Arthur,
        c. 1870
    James Conolly 1985/P/86
    William Clemo 1985/P/87
    Richard Cobbett 1985/P/93
    Charles Downes 1985/P/92
    Nathan Hunt 1985/P/73
    John Merchant 1985/P/72
    Queen Victoria Museum and Art
        Gallery, Launceston
Page 222
    Melbourne Morning Herald,
        11 November 1850
    La Trobe Collection
    State Library of Victoria
Page 226
    Qantas First Aircraft—AVRO 504K
    Qantas Airways Ltd
Page 230
    First Rail excursion, departing Launceston
        to the 6½ mile peg
    from *Illustrated Australian News*,
        4 September 1869
    National Library of Australia
Page 231
    top
    A miniature of Mary Reibey
    ML ref ZMIN 76
    Mitchell Library
    State Library of New South Wales
    bottom
    Photograph of George Reid
    Australian Natives' Association
Page 233
    Tom Roberts   1856–1931
        Australian
    Mrs L. A. Abrahams 1888
    Oil on canvas
    40.8 x 35.9 cm
    Purchased 1946
    National Gallery of Victoria
Page 234
    John Eyre
    View of Sydney from West Side of Cove
    from David Mann, *The Present Picture of*
        New South Wales, 1811
Page 235
    Ellis Rowan   1848–1922
    Unidentified Australian Flowers
    Watercolour
    54.5 x 37.5 cm
    Ellis Rowan Australian Collection
    275/866
    National Library of Australia

Illustrations and Acknowledgements

Page 236
Ron Tandberg
I don't want to go ... but someone's got to protect little countries if they've got lots of oil
H92.163/1
La Trobe Collection
State Library of Victoria

Page 238
A sketch of Bligh's arrest
ML ref Z Safe 4/5
Mitchell Library
State Library of New South Wales

Page 240
Photograph of J. H. Scullin
National Library of Australia

Page 241
Photograph of Dominion League of Western Australia, Grand Victory Festival, 1933
National Library of Australia

Page 242
What's your excuse for not joining your comrades in the AIF?
Tasmaniana Library
State Library of Tasmania

Page 245
Julian Rossi Ashton 1851–1942 Australian
The Selector's Daughter 1907
Watercolour and pencil
53.5 x 36.6 cm
Purchased with the assistance of a special grant from the Government of Victoria 1979
National Gallery of Victoria
Reproduced with the permission of Paul del Prat

Page 246
Ball given to the officers of the Confederate Steamer Shenandoah *at Ballarat*
from *Illustrated Australian News*, 23 February 1865
La Trobe Collection
State Library of Victoria

Page 250
Margaret Preston 1875–1963 Australian
Catherine Helen Spence 1911
Oil on canvas
104.7 x 77.5 cm
Gift of the Citizens Committee 1911
Art Gallery of South Australia

Page 251
George Lambert 1873–1930 Australian
The Squatter's Daughter 1923–4
Oil on canvas
61.4 x 90.2 cm
Australian National Gallery, Canberra

Page 253
Photograph of Nellie Stewart
Mina and May Moore Collection Env. 191
La Trobe Collection
State Library of Victoria

Page 254
Photograph of Jessie Street
National Library of Australia

Page 255
William Strutt 1825–1915
Portrait of John Pascoe Fawkner, founder of Melbourne 1851
Oil on canvas
61.3 x 51.2 cm
National Library of Australia

Page 259
Lionel Lindsay 1874–1961 Australian
Swagman Bending to Lift a Billy 1936
Pen and brown ink
23.7 x 15.6 cm
Presented by the artist 1953
National Gallery of Victoria
Reproduced with the permission of Peter Lindsay

Page 260
Melbourne's Bathing Season Commences
from *Australasian*, 1 December 1923
La Trobe Collection
State Library of Victoria

Page 261
City Improvements in Sydney: Old Buildings Ordered for Demolition
from *Australasian Sketcher*, 31 July 1880
La Trobe Collection
State Library of Victoria

Page 262
Roland Wakelin 1887–1971 Australian
The Bridge Under Construction c. 1928–9
Oil on canvas on board
101.2 x 121.6 cm
Purchased 1967
National Gallery of Victoria

Page 263
A caricature of David Syme by Phil May
David Syme & Co. Ltd

Page 266
A map 1644 by Abel Tasman
ML ref MT3/800/1644/1, 1a
Mitchell Library
State Library of New South Wales

Page 267
Holidays in Southern Tasmania
from *Tasmanian Mail*, 22 December 1906
Archives Office of Tasmania

Page 268
Lottery ticket
Tattersall Sweep Consultation

Page 271
Ticket of Leave Passport
National Library of Australia

Page 273
Unknown
Opening of Cathedral, Townsville 1871
Albumen silver photograph
15.5 x 20.1 cm
Presented by Tim Stranks 1990
National Gallery of Victoria

Page 274
James Northfield
Yesterday Cobb & Co.—To-day TAA
Poster
H15364
La Trobe Collection
State Library of Victoria

Page 275
Anti-transportation poster, 1850
CO280/264
Public Record Office, Kew, England

Page 277
Thomas Cook
Truggernana, Native of the Southern Part of Van Diemen's Land
Watercolour
29.2 x 22 cm
Purchased 1949
Tasmanian Museum and Art Gallery

Page 280
Photograph of agricultural science students at the University of Melbourne taken during the First World War
Author's Collection

Page 284
Mothers in Mourning
Anti-war pamphlet, 1970
Riley Collection
La Trobe Collection
State Library of Victoria

Page 290
William Duke
The Rounding
W. L. Crowther Library
State Library of Tasmania

Page 291
Photograph of Patrick White taken by Eva Pavlovic
Eva Pavlovic

Page 293
Photograph of Gough Whitlam taken by J. Crowther
Australian Information Service

Page 294
Poster for *Struck Oil*
Troedel Collection
CB 4/28
La Trobe Collection
State Library of Victoria

Illustrations and Acknowledgements

Page 295
Delegates at the Annual Conference of Woman's Christian Temperance Union
from *Weekly Times*, 3 December 1898
H42.542/6
La Trobe Collection
State Library of Victoria

Page 298
Rudolph W. W. Koch
Tasmanian Smelting Company Limited, General View of Works
from *Journals and Printed Papers of the Parliament of Tasmania*, vol. XLI, paper no. 69 of 1899, 'Report of the Secretary for Mines for 1898–1899'
La Trobe Collection
State Library of Victoria

Colour Illustrations

J. Northfield
Canberra Federal Capital & Garden City Australia
Poster
H90.105/12
La Trobe Collection
State Library of Victoria

C. Dudley Wood
Seagulls Over Eastern Coastline   1938
The United Commercial Travellers' Association Collection
The University of Melbourne Archives

Moody
Welcome to South Australia's Centenary Celebrations of 1936
Weetman Travel Posters Australia 14
La Trobe Collection
State Library of Victoria

Antoine Phelippeaux   1767–c. 1830
Tableau des decouvertes du Capne. Cook & de la Perouse
Hand-coloured engraving
45.5 x 53.1 cm
National Library of Australia

William Barak   1824–1903   Australian
Corroboree
Earth pigments, wash & pencil traces
56.8 x 80.9 cm
Purchased 1962
National Gallery of Victoria

Eugen von Guérard
Tower Hill   1855
Oil on canvas
68.6 x 122.0 cm
Presented by Mrs E. Thornton to the Department of Conservation and Environment 1966
On loan to Warrnambool Art Gallery since 1978

Map of Victoria or the Port Phillip District, c. 1851, drawn and engraved by J. Rapkin, published by John Tallis & Co., London and New York
Frank and Una Lyons Collection

Illuminated address presented to Charles Alfred Topp by the Premier of Victoria, 1899
Professor C. E. Moorehouse Collection
University of Melbourne Archives

Cover Illustration

Theodore King   active 1855
Australian
The First Parliamentary Election, Bendigo, 1855
Oil on paper on board
32.4 x 47.8 cm
Gift of Mr J. S. Dethbridge 1894
Bendigo Art Gallery

I should like to thank all of the individuals and organizations mentioned above for allowing me to reproduce these works. Every effort has been made to identify copyright holders. Any omissions or errors should be drawn to the attention of the publishers for rectification in any future editions.

I also wish to thank Olga Abrahams, John Bangsund, Janette Bomford, Gary Boulter, Sonja Chalmers, Geraldine Corridon, Jill Davies, Andrew Demetriou, Garry Disher, David Elder, Jim Ellison, Sue Ellison, Dr Charlie Fox, Dr Robin Gerster, Rosemarie Kiss, Frank Lyons, Una Lyons, Peter MacFie, Jo McMillan, Dr Ross McMullin, Nan McNab, Neil McPhee, Tim Mahar, Alan Mason, Sandra Nobes, Dr Carolyn Rasmussen, Professor Henry Reynolds, Peter Rose, Professor A. G. L. Shaw, and Dr David Stockley for their much-appreciated assistance. My greatest debt is to the late Dr Lloyd Robson for his support and encouragement.